BIM in Sr
Practices

Illustrated Case Studies

Edited by Robert Klaschka

Published by NBS

Published by NBS, part of RIBA Enterprises Ltd, The Old Post Office,
St Nicholas Street, Newcastle upon Tyne NE1 1RH

ISBN 978 1 85946 499 1
Stock code 80471

British Library Cataloguing in Publications Data
A catalogue record for this book is available from the British Library.

Commissioning Editor: Sarah Busby
Editor: Sara Hulse
Design: Philip Handley
Typesetting and Production: Michèle Woodger and Philip Handley
Printed and bound by W&G Baird Ltd in Great Britain

NBS is part of RIBA Enterprises Ltd
www.ribaenterprises.com

Preface

Small practices have always been innovators. This book is the story of ten innovators in the use of BIM. It celebrates sole traders who are making the most of 'lonely BIM' to improve their productivity and branching out into working with others. It features young and growing practices who have begun working with other disciplines in truly collaborative BIM. Lastly it explores the work of the specialists who have developed specific processes to complement their workflows such as Architype and Eurobuild with their semi-automated Passivhaus BIM systems.

The case studies are supplemented by a series of contextual chapters written by industry leaders and discussing key issues of interest to small practices. Ray Crotty answers the question 'Why BIM?', exploring how our world will change when technology, standardisation and processes used in other industries cross over into design and construction. David Miller discusses the costs, practicalities and benefits of moving your practice to take advantage of BIM workflows and technologies. Stefan Mordue and Rebecca De Cicco give us a rundown of the vast and diverse resources available to those practices just starting to use BIM, including free guidance, groups, training and events. Finally I discuss how BIM can open up other lines for work through diversification to give your business a stable baseline cash flow.

The change to model-based data-rich working to produce design, construction and maintenance information about buildings is probably the biggest shake-up of the way we work that I will experience in at least the first half of my career. The influence of the changes that are being demanded by the government and by private companies who are seeking reductions in construction costs and better data on and control of their buildings is far-reaching and it is clear that these changes will have a lasting effect on both big and small practices. With the UK establishing best practice for collaborative BIM working in the 1192 suite of documents and in future building control sign-off from models, it is only a matter of time before the BIM revolution affects us in one way or another.

Looking more widely, the government construction strategy seeks to consolidate the UK's position of excellence as a design, construction and maintenance services exporter. Through this BIM presents a huge opportunity for versatile small practices who want to grow into this new digital world. With government's commitment to give 25% of all public work to SMEs and ongoing funded research programmes by the Technology Strategy Board specifically focused on digital construction there has never been a better time to start using BIM.

This book highlights some of the best work that small practices are doing in BIM for both themselves and their clients through a series of detailed case studies. It seeks to dispel the myth that BIM is not for SMEs and presents a wide range of examples of how technology and processes can be tailored to work well in a small office to serve high-quality design and drive business success. Most of all the case studies show that small practices can benefit from BIM. The hurdles for entry are less difficult than you think and, more importantly, BIM presents a platform for the future development of your practice as a provider of both exceptional design and valuable information.

Robert Klaschka
June 2014

Foreword by John Eynon

This book busts a few myths about BIM and also, I think, points the way for the future of the profession.

Myth 1 – BIM is for large projects. No! Examples in this book include residential conversions, extensions and a self-build house.

Myth 2 – BIM is only for new buildings. No! There are lots of examples here of work involving existing buildings at all stages.

Let's remember that the majority of practices in the UK are micro to small and medium-sized enterprises. These businesses are the engine room of the design economy and the lifeblood of most projects. These case studies from hands-on practices illustrate and explain the use of BIM right at the grass roots, providing practical examples and plain speaking advice for any practitioner at whatever point on your BIM journey.

Importantly for the architectural profession some common themes emerge here:

- improved collaboration for the team
- better design and project coordination
- increased efficiency and profitability
- more value added services; more accurate information earlier in the process
- better coordination with contractors and subcontractors
- faster and more accurate optioneering of alternative solutions at any stage.

Architect SMEs are the speedboats of the industry, weaving between their oil tanker rivals and counterparts – they are quick to adapt, innovate and evolve with market conditions. BIM provides an opportunity to regain ground in design, procurement and construction processes by managing, coordinating and delivering perfect data across the asset life cycle and cementing their place as the real master builders in the industry.

I commend this book to all architects – this is required reading for anyone thinking about starting their BIM journey or already on the road.

Carpe diem!

John Eynon is a chartered architect and construction manager, writer and presenter, Chair of the SEBIMHUB and Vice Chair of BIM4SME, a CIOB Ambassador and Director of Open Water Consulting.

Foreword by David Philp

Building information modelling (BIM) is emerging as an enabler of more efficient, innovative and collaborative forms of working. This value proposition is especially relevant to the smaller players in the built environment who are more agile. Whilst for some there may be a need to make an initial investment, the realisation of quicker workflows, automation of key processes and better coordination of information will soon unlock a positive return on their outlay.

We are also witnessing some early SME adopters of BIM punching well above their weight and levelling the playing field against their bigger competitors, offering added value, less risk and better outcomes. As you will see through the case studies in this book, BIM, and in particular virtual prototyping, is helping practices and their customers to better understand design solutions as well as ultimately improving the quality of builds. From better spatial coordination, rapid optioneering to self-generation of data sheets the list of opportunities for the small practice is a large one. Organisations must also consider the risk of doing nothing in a sector that is rapidly shifting to digital transactions and queries.

Hopefully this book will inspire you to make changes in your organisation – to either start doing BIM or perhaps accelerate your implementation.
To paraphrase Clint Eastwood, "Sometimes if you want to see a change for the better, you have to take things into your own hands."

Professor David Philp, MSC, BSc, FRICS, FCIOB, FGBC
Head of BIM – UK BIM Task Group

Acknowledgements

This book is dedicated to my wife Philippa, my son George and Izzy, aka BIMdog.

I'd also like to thank Mervyn Richards and Nick Nisbet of BuildingSMART. Their ideas, passion and determination have inspired my journey into the BIM world and regularly remind me that it is not a new idea but one that has been maturing for decades.

Books like this can't happen without the involvement of many others, not unlike successful BIM. Beyond the hard work that the chapter authors have all put into their contributions, we've had the help of a distinguished group of reviewers supporting us. My thanks to David Philp, John Eynon, Richard Fairhead, Paul Vonberg and Rachael Davidson.

Thank you also to Michèle Woodger and Philip Handley for their fantastic work on the production and design of this book.

Last but certainly not least my thanks go to Sarah Busby, my long suffering keeper at RIBA Publishing. Without her constant support, advice and direction throughout the creation of the book, it would never have happened.

It seems appropriate, given that this book crosses many of the authoring and sharing platforms involved in the BIM process, that it has been supported by most of the companies whose software features in these pages. My thanks go to Autodesk, Bentley, Solibri and Vectorworks for their support.

Robert Klaschka
June 2014

Contents

Notes from our publishing partners vi-ix

▶ PART A **Case studies 01–10**

Case study matrix 1

01 BIM and the new RIBA Plan of Work 2013 — Poulter Architects 3

02 Communicating Ideas with BIM — Jonathan Reeves Architecture 11

03 BIM and Clients — jhd Architects 19

04 Traditional and Modern Methods of Construction using BIM — Axis Design Architects Ltd 27

05 Winning Funding with BIM — Constructive Thinking Studio Ltd 35

06 Moving Towards BIM in the Cloud — Niven Architects 43

07 BIM for Existing Buildings — Studio Klaschka 51

08 BIM for Passivhaus Design — Architype Ltd 59

09 Building Passivhaus with BIM — Eurobuild 67

10 BIM and Working with Consultants — Metz Architects 73

▶ PART B **Contextual chapters 01–04**

01 Why BIM? — Ray Crotty 83

02 Getting Started with BIM — David Miller 87

03 Tapping into the BIM Community — Stefan Mordue and Rebecca De Cicco 89

04 BIM and Diversification — Robert Klaschka 91

BIM Glossary — Rob Jackson 93

Publishing partners

Autodesk helps people imagine, design and create a better world. Customers across the manufacturing, architecture, building, construction and media and entertainment industries – including the last 19 Academy Award winners for Best Visual Effects – use Autodesk software to design, visualise and simulate their ideas before they're ever built or created.

Within the infrastructure and construction industries, BIM is changing how buildings, infrastructure and utilities are planned, designed, built and managed. Autodesk BIM solutions help turn information into insight and deliver business value at every step in the process. BIM gets the right information to the right people at the right time, helping firms innovate and compete. Autodesk's comprehensive BIM portfolio delivers a workflow advantage for virtually any project.

Publishing partners

Bentley Systems has strong BIM credentials with ability to support companies working collaboratively through the life cycle of a project. Bentley supplies products supporting information modelling, management and mobility and emphasises how federated models can provide more efficient collaboration in a multidisciplinary project team BIM workflow. Fully supportive of IFC and COBie, AECOsim Building Designer, Bentley's interdisciplinary BIM application, provides COBie-compliant IFC (SPFF) files and COBie spreadsheets (from Bentley i-models).

Publishing partners

In the BIM world we are creating more data than ever before. This brings with it the real probability that, if this data is not properly verified and managed, potential problems currently experienced in traditional construction processes will be magnified.

The answer is to provide a solution that can assure data quality at both the single and combined (federated) model level. Solibri Model Checker provides that solution.

Using a unique rules-based system, models can be assessed to determine if they comply with generic construction requirements such as design coordination, clash detection and constructability analysis, as well as specific requirements such as correct and complete COBie data, Building Regulations compliance and meeting bespoke design specification.

Having ensured that designs can actually be built (a massive leap forward from traditional process) Solibri Model Checker can then be applied to interrogate models to provide key information to users, such as quantity take off, COBie spreadsheets, product data and space boundaries.

Imagine – reliable information on demand. With Solibri Model Checker this becomes a reality, and for that reason it is the product of choice for the construction professional.

Publishing partners

For 25+ years, Nemetschek Vectorworks Inc has been a global leader in design technologies, providing elegant architectural and landscape design software that offers BIM capabilities in a flexible, hybrid-design environment. As members of BuildingSMART, we are committed to 'openBIM' and true interoperability within this industry. These open standards include the use of Industry Foundation Classes (IFC) for building geometry and data. We are actively responding to the UK government's BIM mandate and COBie compatibility, with particular attention to the needs of small and medium-sized practices and also take an active role within the Landscape Institute BIM Working Group and BALI Affiliate Members BIM Committee.

Over the last few years, Nemetschek Vectorworks has continued to raise its profile as a major BIM solution, producing BIM guides and running RIBA CPD certified BIM seminars. We also partner with various industry companies, such as NBS, to provide collaborative BIM workflows, an example being the Vectorworks NBS Annotator plug-in. We are also one of the founder content providers for the National BIM Library.

About the editor

Robert Klaschka has over 15 years' experience working in the UK and Europe in architecture and design and is founding director of Studio Klaschka, a small architecture practice specialising in retrofitting and refurbishment, based in South London. Over the ten years of the practice, he has driven the adoption and use of BIM software to make the most of his small team's resources. He has experience in cultural, developer-residential, student accommodation, sports and office fit-out projects. He has close links with the development team at Bentley Systems and was elected chair of The Bentley Community UK and Ireland in 2010 and 2014. As an Advisory Panel member for NBS he has advised on the development of the data exchange between NBS Building and Bentley Architecture. Whilst there is no question that he enjoys working at the leading edge of design software use and development, Robert's focus is on maximising design flexibility and creativity, whilst understanding the consequences of design and briefing decisions.

Part A: Case Studies

01	BIM and the new RIBA Plan of Work 2013 Poulter Architects	3
02	Communicating Ideas with BIM Jonathan Reeves Architecture	11
03	BIM and Clients jhd Architects	19
04	Traditional and Modern Methods of Construction using BIM Axis Design Architects Ltd	27
05	Winning Funding with BIM Constructive Thinking Studio Ltd	35
06	Moving towards BIM in the Cloud Niven Architects	43
07	BIM for Existing Buildings Studio Klaschka	51
08	BIM for Passivhaus Design Architype Ltd	59
09	Building Passivhaus with BIM Eurobuild	67
10	BIM and Working with Consultants Metz Architects	73

CASE STUDY MATRIX

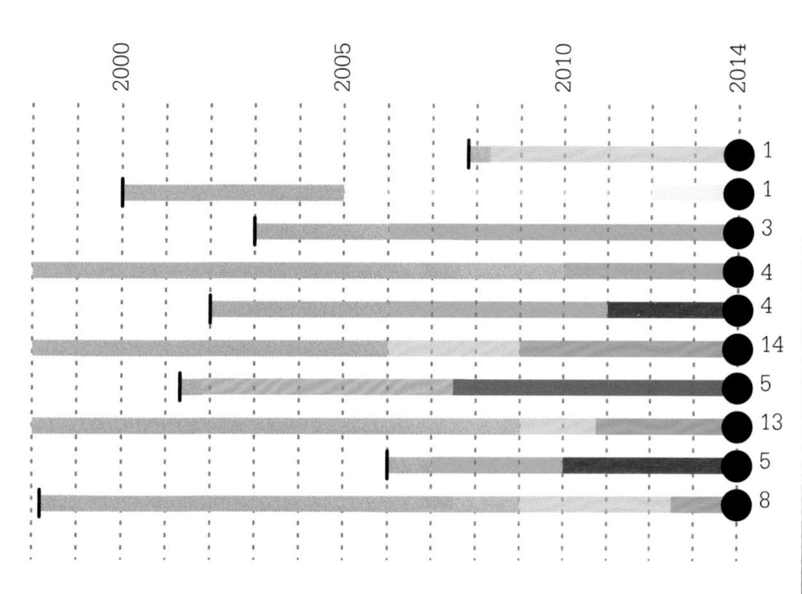

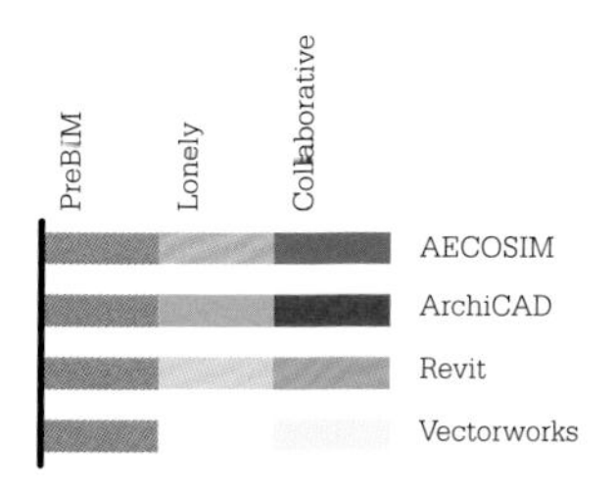

BIM AND THE NEW RIBA PLAN OF WORK 2013

Jeremy Poulter

Practice:
Poulter Architects

Location:
New Mills, High Peak, Derbyshire

Founder:
Jeremy Poulter

Employees:
Sole trader

Founded:
2010

Technology:
Autodesk Architecture Suite 2011, including Revit and AutoCAD; Adobe Photoshop; Microsoft Office

01

This case study demonstrates how a sole practitioner is able to utilise the benefits of BIM to his own advantage on the projects that are his bread and butter – residential projects. It also explores how, even in the smallest of practices, the BIM process can be mapped to the stages of the new RIBA Plan of Work 2013.

'The more information you start producing, the greater the benefit of a BIM object over a traditional 2D file.'

THE PRACTICE

One of my first investments when I set up the practice was BIM software, in the form of Autodesk Revit. Whilst working at Bond Bryan Architects (until the LSC funding fiasco left me out of work), I gained some experience of ArchiCAD, a Mac-based BIM package and saw the potential of BIM to transform the way we produce and communicate building design.

The bulk of my experience, however, has been using AutoCAD on a PC. I chose the Autodesk product Revit over ArchiCAD, out of familiarity with the PC environment and because, within the industry, it looked like it would become the most popular BIM package.

The majority of the practice's projects have been residential extensions, renovations and new-builds for private clients.

All projects are modelled in 3D in Revit after an initial hand-drawn sketch design stage. This often involves the modelling of existing buildings and landscapes.

From the model we generate:

- 2D drawings: plans, sections, elevations, details
- 3D drawings: axonometrics, line perspectives, rendered perspectives, solar studies
- Schedules: Area, Door and Window, Materials, Drawings.

The other consultants we work with all work in a traditional 2D format. We model their designs in 3D within the Revit model and set them out. This includes all the main elements of structure and significant building services elements. Any information we export to consultants or specialist subcontractors are 2D CAD files generated from the Revit model.

This is very much Level 1 or 'lonely BIM', where only one member of the project team is benefitting from the model.

However, given the scale of projects that we tend to work on, it does mean we are able to have full control over the management and coordination of the design in one single model.

THE PROJECT

▶ 2 ST PETER'S AVENUE, KNUTSFORD

This case study tells the story of how I have used BIM on one residential project from its inception all the way through to its completion. The latest RIBA Plan of Work stages are used as broad headings, providing an indication as to where we were in the process and how the model was used to progress through each stage.

■ **Stage 0** Strategic definition

The client, who is a surveyor, was introduced through a mutual acquaintance late in 2011. He was in the process of purchasing an existing two bedroom bungalow within walking distance of Knutsford town centre. He had identified its potential for extensive renovation and enlargement into a low-carbon family home. He had also already engaged a structural engineer who established, in principle, that the existing footings would allow an additional storey to be built on top of the existing structure.

He initiated a mini competition and asked both Poulter Architects and another practice to produce initial sketch designs based on an initial brief and the dimensioned sales literature.

We produced an outline 3D model in Revit from which sketch options were developed. The hand drawn and photoshopped perspectives we presented were tracings of the 3D perspectives generated from the Revit model.

On the basis of the initial options and our presentation Poulter Architects were appointed for the project.

Figure 1.1
The existing bungalow

■ **Stage 1** Preparation and brief

In terms of procurement, the client had decided to project manage the building himself and would be procuring trades to carry out the various elements of the build. Our scope was to take the project through planning, produce a full set of working drawings and carry out a number of site visits to monitor progress.

I undertook a measured survey of the property and plot and then modelled this in 3D in Revit. I was then able to add information on the depth and size of the footings and had gleaned some information about the existing wall constructions which I was able to add to the model.

From this I produced a set of 2D survey drawings: for some of the existing elements it was simpler to draw things in 2D. Below-ground drainage, for example, was simply an annotated mark-up of the site plan based upon the measured manhole positions, invert levels and a CCTV survey of drains.

■ **Stage 2** Concept design

Having already carried out the sketch options we discussed and agreed which elements of each were to be combined and developed during Stage 2. We also obtained some pre-application advice from the planners, who highlighted the need to address the building's relationship with the conservation area neighbouring the site.

A key driver in the design of the building was the location and configuration of the stairs and the desire to get a view down into the living room from either the first floor or the staircase. Several options were explored, modelled in 3D, then 2D drawings were generated.

The benefit of using the model was that the implications of each could be explored more quickly than going through the process of drawing out both plans and sections for each option to see if something would work in the way we wanted it to.

Figure 1.2
Sketch options which used perspectives generated in Revit as a base

For example, we initially looked at a solution where a first floor gallery overlooked the living space. What became apparent quite quickly was that the roof to the living room would need to be lifted quite significantly to make this work, such that it would make the relationship of the single-storey and two-storey elements seem out of proportion.

The solution we settled on placed the stair in the middle of the plan, so that the half-landing provides a viewing point overlooking the living room and allows views both up and down from the living space.

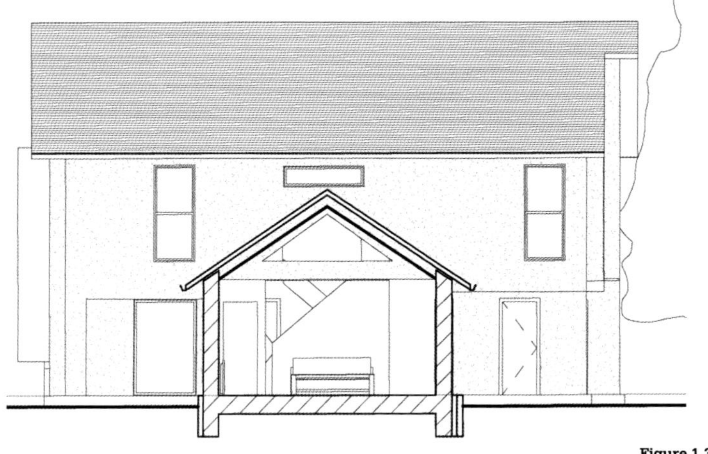

Figure 1.3
Early section generated in Revit demonstrates limited visual connection between first floor and living room

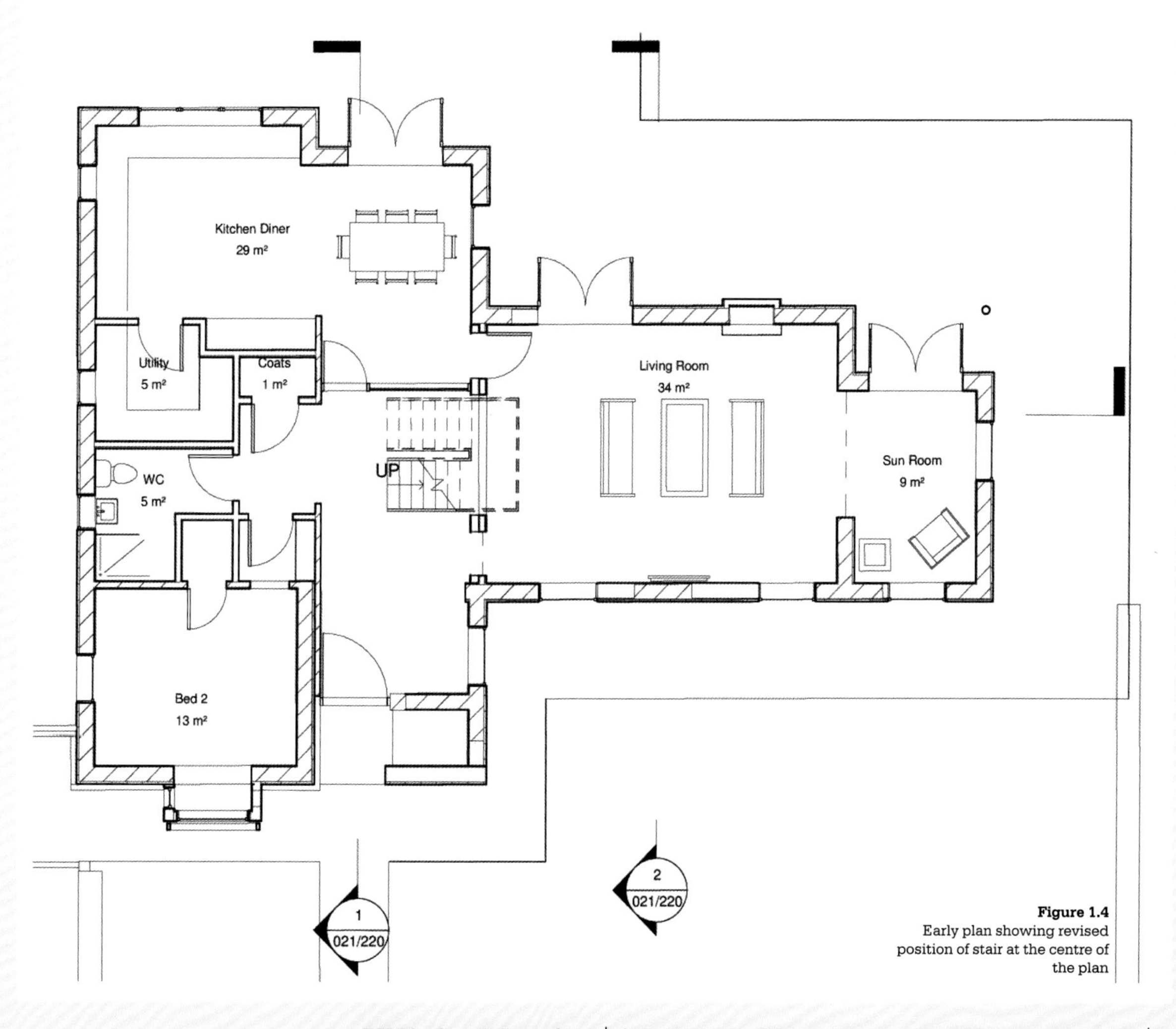

Figure 1.4
Early plan showing revised position of stair at the centre of the plan

■ Stage 3 Developed design

Once the design was frozen, information was developed towards a planning application. One of the issues with BIM is that, at this stage, you find you put more information into the model than is necessarily apparent in the drawings. A lot of the general arrangement (GA) coordination occurs during this stage and modelling in 3D forces you to put additional effort into resolving the tricky junctions. The huge advantage over designing in 2D is that coordination does not rely on the designer following through any design decision made in plan through the sections and elevations.

Decisions were also taken about the build-up of walls, floors, etc. and fed into the model even though they are not seen in the 2D drawings. Walls were drawn as a single composite entity and not the series of lines you would draw in 2D. A handy detail level button simply switches the wall construction from two lines, showing overall wall thickness, to the composite build-up of construction layers, ready for Stage 4.

The other great benefit of Revit is the ability to put materials, colour and shadow on surfaces. This was used in the elevation drawings for planning. The finish for each wall is selected when you create the wall construction and when viewed in elevation the colour and texture can be turned on. Shadows can also be turned on and set to any time of day or any date in the year. A fully coloured and shaded elevation can be created without the laborious process of colouring up by hand or in Photoshop. Even better is that when you adjust the design, shadows and colour instantly update in the elevations, preventing the scenario I used to see very often, of a design change leading to many hours of work recoloring or photoshopping drawings.

Rendered perspectives were also generated in Revit to show how the building would look both internally and externally. Again the material finish was simply allocated to the wall, floor, etc. within its properties dialogue.

Figure 1.5
Elevation submitted for planning showing colour, texture and shadow applied in Revit

Perspectives were simply set up and rendered without the need to export to other software and then carry out further work on allocating materials and lighting. Whilst the renders may not be the polished photorealistic images achievable in other software, for the scale and type of project I work on, they provide a great means of communicating my designs.

Phasing issues also needed to be looked at in greater detail at this stage and the software helps to address this. The basis of the whole model is the existing building which is allocated to the Existing Phase. In the New Construction Phase existing elements (walls, floors, roofs, windows, etc.) can either be retained or demolished. You can simply switch between phases in any view to understand what is being retained or demolished.

Figure 1.6
Internal rendering of the living room looking towards the stair

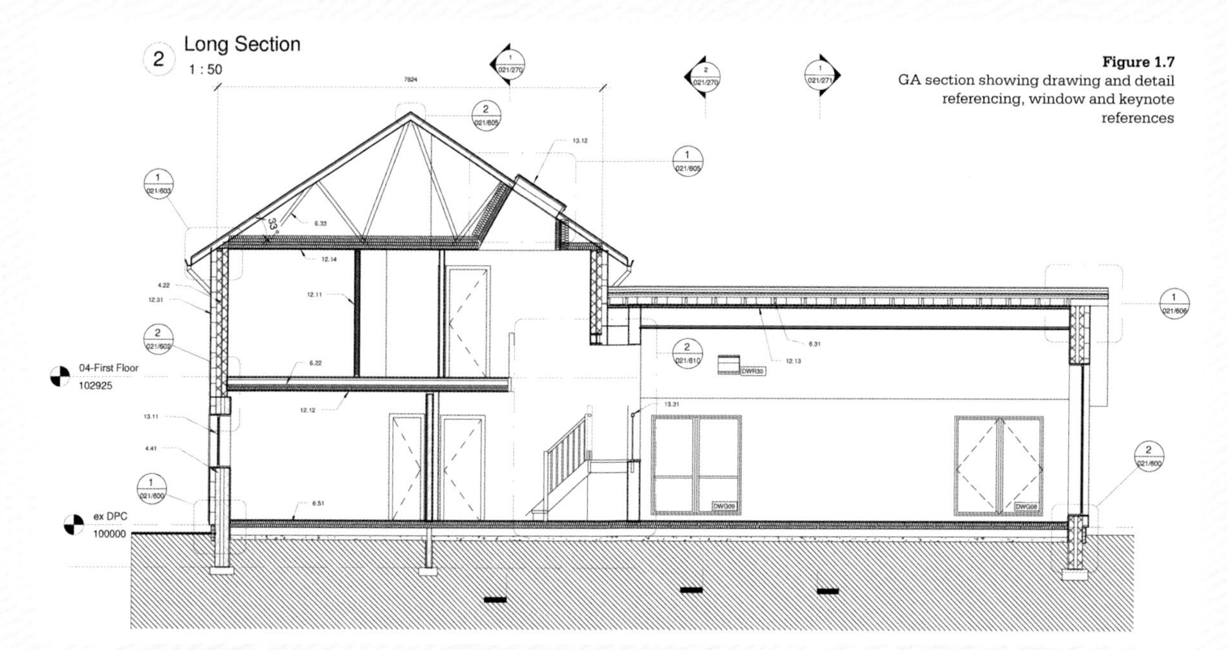

Figure 1.7
GA section showing drawing and detail referencing, window and keynote references

Because this project required a substantial amount of demolition, this caused issues in Revit, as new knock-through openings didn't seem to be able to be allocated to a phase. This meant that both Existing and New Construction model files were required rather than the ideal situation of a single model file.

■ **Stage 4** Technical design

The more information you start producing, the greater the benefit of a building information model over traditional 2D files. Nowhere is this more apparent than in Stage 4.

I had already built in a lot of information about wall, floor and roof constructions in Stage 3 and by switching the detail level from Coarse to Fine, the composite build-up of these elements now appeared in the GA drawings.

From here several more GA sections were set up and allocated to sheets. At this point it required very little additional work to get these basic sections. Key details were identified and again set up on sheets, very quickly setting up a drawing schedule and structure.

A great benefit of the Revit model is that it references sections and key details from the GA to the drawing sheet.

There is no need to draw and create your own detail or section reference. If you change the drawing number the reference updates through the model and vice versa. This creates a robust drawing management structure.

The sections and key details that are set up are basic outlines showing the build-up and junction of elements. The bulk of the work then is in adding 2D detail, annotations and dimensioning to these outlines to resolve these junctions.

At this stage the engineer sized all of the structure and I modelled this information, positioned and set out these elements. 2D CAD files were then exported out of the model, which the engineer used to produce a set of calculations to back up the structural design.

The specification was a separate stand-alone Word document, referenced on the details by material keynotes. A keynote is a reference you give a material or element within Revit that labels selected materials automatically, rather than having to type in each label by hand. This does require some upfront work, but saves a significant amount of time spent working out which label goes with which material on each and every drawing.

Following a review of costs, there were some changes to the design. For example we originally modelled the front brick entrance bay in a traditional cavity wall construction. However from a cost and construction perspective it was decided it would be easier to use the same external wall insulation (EWI) system cloaking the rest of the building, but with a stick-on brick instead of render.

The reduction in man-hours to make this change, when compared with a 2D approach, is significant. In terms of the GAs it was as simple as changing the composition of the wall construction whilst maintaining the external setting out points, rather than having to go through each plan and section to make the changes. However the revised details did require reworking in 2D, which took some time.

The door and window schedules were also generated in Revit with door and window references indicated on the plans, linked to the schedule, thus avoiding the laborious exercise of creating a stand-alone schedule. Overall during this stage, 26 A2 drawing sheets were produced, plus a specification. This was approximately five weeks of work in total. It is true that a great deal of coordination was done in Stage 3, that in a 2D set-up may well

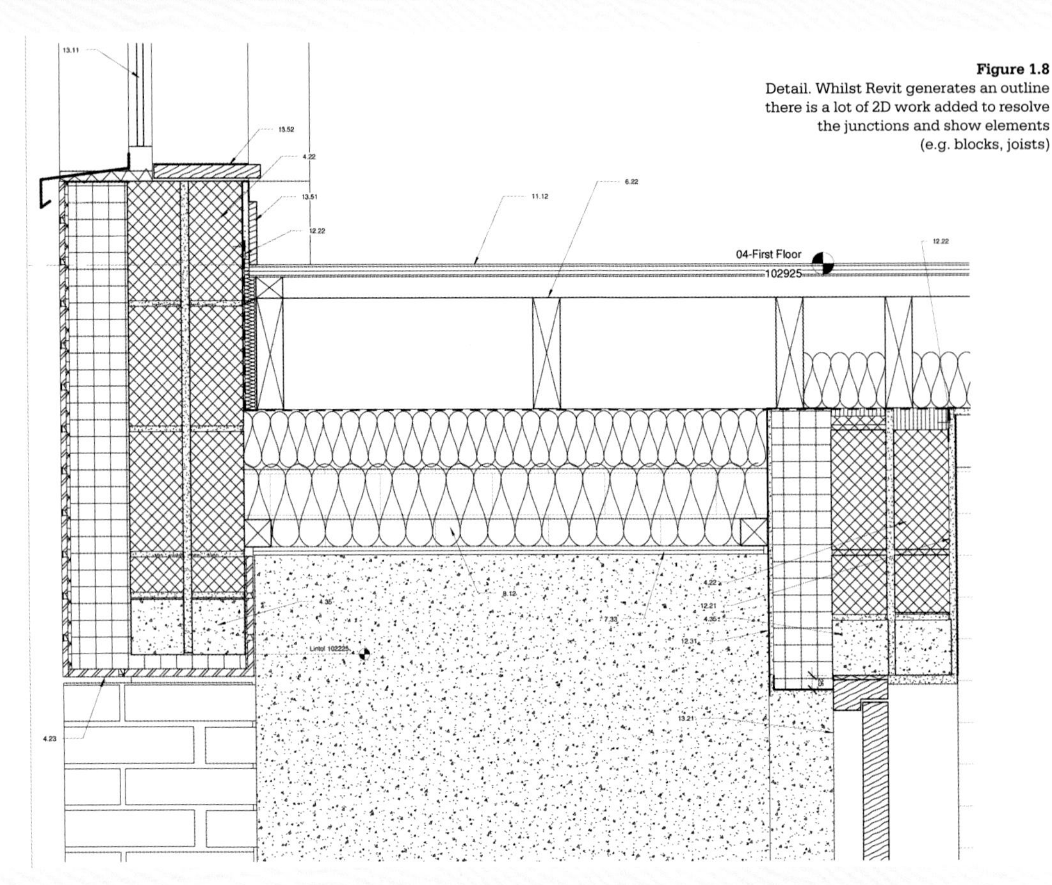

Figure 1.8
Detail. Whilst Revit generates an outline there is a lot of 2D work added to resolve the junctions and show elements (e.g. blocks, joists)

'The benefits of a fully coordinated design, which have been assisted by modelling the project using BIM software, means we have avoided any uncoordinated junctions and having to design our way out of these.'

have been done ordinarily in Stage 4, but from experience I know this is a considerable improvement on what I would have been able to achieve, had the work been carried out in 2D. The other great benefit of modelling in 3D is the confidence you have in the coordination of the design.

▪ **Stage 5** Construction

Work started on-site in December 2012 and I visited site to monitor progress roughly every other month throughout the build.

The biggest challenge has been the variance in the level of the existing brick coursing, which only came to light once the existing internal plaster was removed. The Revit model was very helpful in terms of identifying a datum level and making the necessary adjustments to the coursing.

The most significant design change has been the client's decision to demolish the existing suspended timber floor and instead install a ground-bearing slab. There has been an adjustment to the detailing but the finished floor level remains as it was in the original design.

Otherwise, there have been very few site queries and the work has progressed well. The benefits of a fully coordinated design, which have been assisted by modelling the project using BIM software, means we have avoided any uncoordinated junctions and having to design our way out of these.

CONCLUSIONS AND FUTURE PLANS

Integrating BIM into the way the practice works, right from the very start, has been a huge benefit to its development and what it has been able to achieve. At the very beginning, when there was very little work, time was spent experimenting with the software and discovering its capabilities. This was time well spent, because when work did start coming through the door, I was well prepared.

When compared with a 2D CAD approach, of which I had many years' experience, BIM has allowed me to deliver more, in less time and better communicate the proposals to clients through the drawings and 3D images the model is able to generate. On relatively small projects, such as that discussed in the case study, being in complete control of a single building information model, including coordinating all the other consultants and specialist designs, offers great advantages.

The change to workflow from a 2D approach means that there is perhaps more upfront coordination work in Stage 3 and a more complete coordinated package at the end of Stage 4. The benefit is that during the construction phase there are far fewer site queries and time isn't spent trying to resolve dimensional discrepancies.

In terms of how I see this developing further within the practice, I have already started to carry out cost estimates using the materials schedules and areas generated in Revit combined with information from Spons. I'm now looking at adding material areas to the specification, so that when contractors tender for a project, they use an accurate take-off, leading to a more accurate price and less waste when the project goes to site.

I am also exploring the possibility of purchasing IESVE, an environmental analysis tool to which Revit models can be imported. This is capable of assessing performance requirements such as SAP, SBEM, BREEAM or daylighting, at any stage during the design process. Whilst this will not replace the need to get independent qualified environmental assessors to carry out statutory assessments, it will allow me to make better informed decisions early in the design process regarding issues such as window sizes, solar shading and thermal performance.

Figure 1.9
The building nearing completion

BIM does offer the small practice the opportunity of greater control over the design and construction process through the greater ease in being able to produce fully coordinated and integrated designs. The smaller scale of the projects I work on means that I can combine the roles of architect, lead consultant and BIM manager. It also offers the potential to provide further services such as environmental analysis and materials scheduling.

I am keen to grow the practice and work on larger and more challenging projects. With a good understanding of Level 1 BIM, where I am the sole team member managing the model, the next step is to Level 2, where you work as part of an integrated team of consultants and specialist designers all using BIM as a collaborative tool. When that opportunity arises I hope to make that step up to Level 2.

'...being in complete control of a single building information model, including coordinating all the other consultants and specialist designs, offers great advantages.'

LESSONS LEARNT

My three tips for any practice considering BIM would be:

- Don't be scared of the change. Creating models is actually quite easy to pick up and intuitive for anyone with a good grasp of buildings and 3D.
- Make the change. BIM will lead to a sea change in the way buildings are designed, procured, constructed and operated. Architects have a great opportunity to regain a leading role so get on board before you get left behind. But...
- Don't run before you can walk. Start at Level 1 and gain a full understanding of the capabilities of the software through training and exploration before progressing further.

COMMUNICATING IDEAS WITH BIM

Jonathan Reeves

Practice:
Jonathan Reeves Architecture

Director:
Jonathan Reeves

Location:
Ilfracombe, North Devon

Employees:
1 architect, 1 administrator

Founded:
2000

Technology:
Vectorworks Architect; Cinema 4D; Artlantis; Adobe Photoshop

02

This case focuses on the intuitive nature of communication using 3D models. To some extent we all take 3D visualisation for granted now as it is so prevalent.

In the case study the inherent efficiencies of producing 3D visualisation from design and construction models are discussed. The study discusses solutions that improve your internal productivity and provide the basis of high-quality visualisation models as part of the process.

'As a small practice that often collaborates with other architects and developers, being able to communicate ideas clearly in three dimensions has always been important to us.'

THE PRACTICE

We are a small practice based in the south-west, mainly concentrating on work for private clients and BIM consultancy for other practices. Having always used Vectorworks for 2D and 3D work, our adoption of BIM has been relatively straightforward, as we had already been doing some elements of the BIM workflow before the term BIM became part of modern architectural vocabulary.

Vectorworks Architect is our main tool for all 2D CAD, 3D modelling and also building information modelling. We find it a powerful cross-platform solution that is ideal for small to medium-sized practices and projects. Its ease of use and powerful toolset make it very productive for all the main tasks within the architectural design process, from concept design right through to production information and also marketing brochures.

Those of you familiar with Vectorworks will know that it comprises a number of modules that cover most aspects of building design and construction, from terrain modelling, site and landscape design through to rendering and visualisation tools such as Spotlight for lighting studies and Renderworks for advanced ray traced rendering.

Vectorworks Architect includes free-form modelling capabilities that allow users to design any object from any 3D view including graphics, interiors, hardware, furniture fittings and the site itself – with a healthy crop of trees and plants. In fact we have recently been producing a series of eight garden graphic animations using Vectorworks Landmark combined with Artlantis for the popular ITV show 'Love Your Garden', which shows how flexible a program Vectorworks is for landscape design as well as architecture.

Figure 2.1
Still taken from the 'Love Your Garden' TV animation

As interoperability and collaboration are the essence of effective BIM, Vectorworks includes a range of options, including IFC 2x3 and gb.XML file handling, the ability to exchange 3D NURBS models with Rhinoceros 3D, simple importing of concept models from Sketchup or downloading of thousands of models from Trimble's 3D warehouse. It can also import and export the latest version of AutoCAD and earlier files and export 3D models to Google Earth.

While the definition of what BIM really means is still developing and can mean different things to different people, there is a growing acceptance that 2D CAD drafting using lines and shapes does not exploit the full potential of developments in both software and hardware.

Some architects tend to think that BIM is only suitable for larger projects involving lots of coordination with the entire design team all using the same software. However, the fact that smaller architects are not just one-stop house designers any more and that they need to adhere to building regulations, environmental standards and to satisfy local planning offices, clients and any other parties interested in local development, means that they need to call on experts in each particular area to support their work.

This has been made easier by the ability to create 3D models of their designs and to share these with other people or companies drawn into the project, so that they can view and comment on the plans, analyse the model to see if it meets local sustainable targets, or further the design by adding their own components – structural elements, MEP and so on.

SEEING THE BENEFITS OF BIM AT KINGSWAY SCHOOL

As a small practice that often collaborates with other architects and developers, being able to communicate ideas clearly in three dimensions has always been important to us. As an early adopter of BIM, the first real practical experience of the benefits was collaborating on a project for Kingsway School, helping Quattro Design Architects in Gloucester back in 2006. (This project featured last year in CAD User and Detail 4 2011).

Figure 2.2
Aerial image of Kingsway School showing the play space between the primary and junior school wings

Figure 2.3
Image of Kingsway School courtyard

Figure 2.4
Elevations of Extracare Nursing Home generated from the BIM

The design was a colourful series of angled wings with double curved zinc roofs, which would have been difficult to draw traditionally and visualise by the client teachers. While little collaborative BIM was achieved on this project, the ability to evaluate the design visually, generate accurate plans, sections and elevations and also produce shadow studies and presentation material to help secure funding was invaluable. The project was so successful that Quattro have recently been commissioned to design a very similar sister school a few miles away for the same client. The experiences learnt on the Kingsway project were then applied to a much larger three-storey Extra Care nursing home in Ebbw Vale discussed later.

The design for this project was more complex and the timescale for development very short, so producing a full set of planning drawings from a single coordinated model was an essential part of the workflow. There was also the additional benefit of being able to produce a set of marketing visuals and a full animation moving from external setting to inside journey through the building. This animation can be viewed on YouTube at: the jra-vectorworks-cad channel.

We are now using the latest release of Vectorworks Architect 2014, which has full BIM capabilities on projects of all types and sizes in the south-west of the UK, especially in the counties surrounding our base in Ilfracombe.

PASSING ON OUR KNOWLEDGE

As one of the pioneers of Vectorworks 3D workflows, we now not only provide architectural services, but also offer Vectorworks CAD training, sales and BIM consultation and visualisation services for architects.

We are recognised as one of the leading architectural Vectorworks trainers in the UK and provide both one to one and on-site training courses at all levels to clients big and small. Being respected as a teacher that talks their language and also a successful practicing architect with hands-on BIM experience is the key to our success with practices.

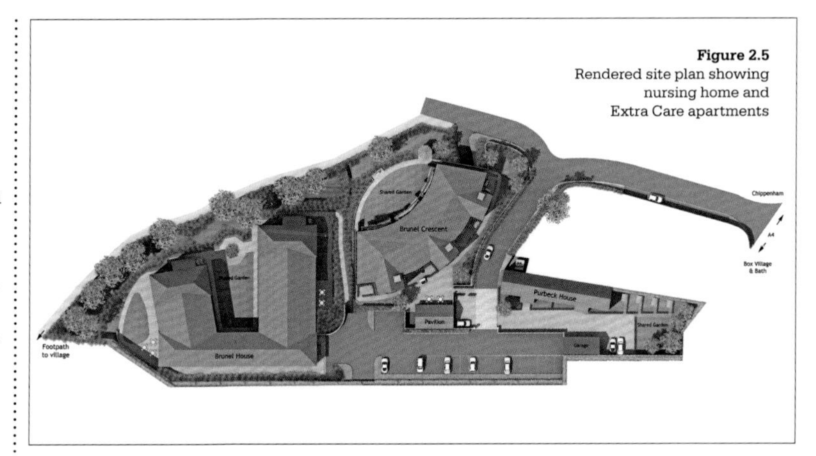

Figure 2.5
Rendered site plan showing nursing home and Extra Care apartments

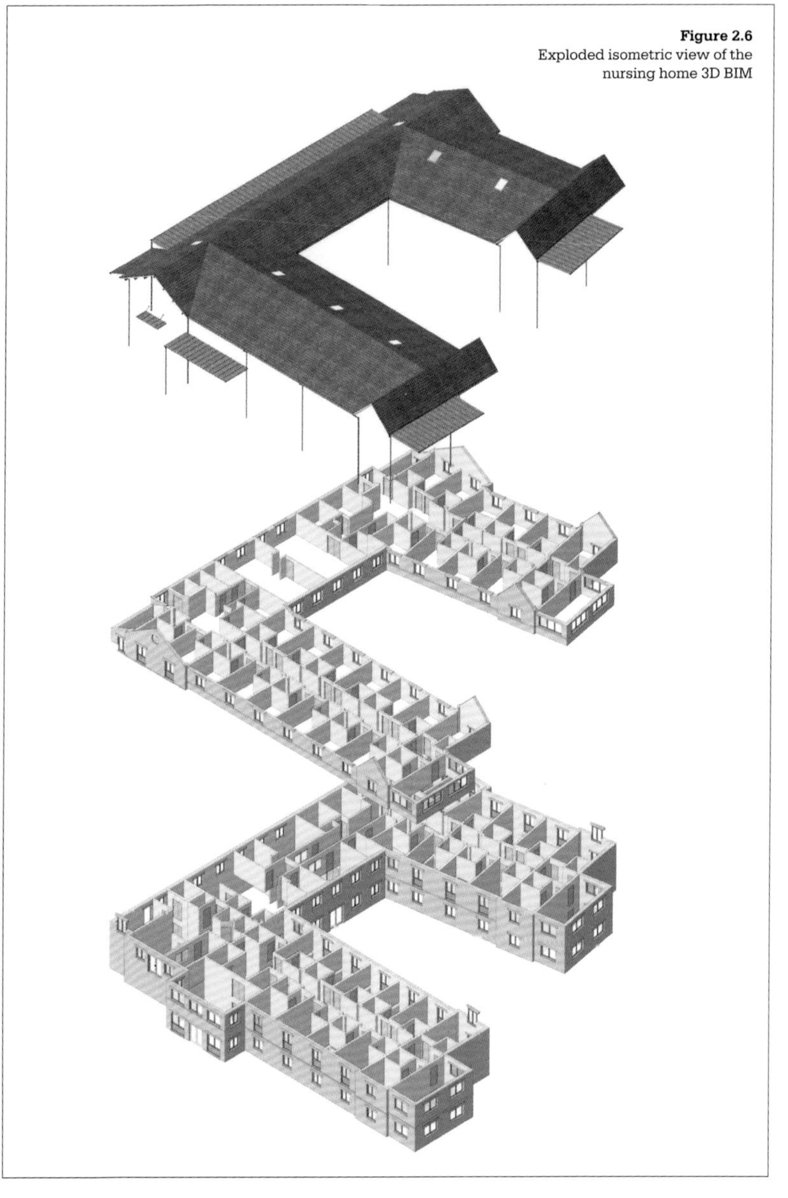

Figure 2.6
Exploded isometric view of the nursing home 3D BIM

Figure 2.7
CGI visual of the development used for project marketing

We find that perceived lack of time, confidence and investment is what holds them back from adopting more BIM into their workflow. However, once they have had some good grounding in the basics they can see the benefits and often feel that there is no going back. That's when they begin to fully engage.

As well as providing a full range of Vectorworks based training courses, we offer a BIM design and modelling service as remote consultants, working with other architects or developers, some of which have been carried out without ever meeting the client. While this may sound strange to some, the advent of fast broadband, powerful laptops capable of running advanced BIM software and ease of communication via Skype, mobile phones and technologies such as FaceTime and instant messaging have changed the requirement to work geographically near to the client.

It was working in this way that we have been involved with a developer and lead architect Llewellyn Harker Architects (LHA) on a number of projects including a church conversion in France and an underground ballroom for a listed building in Box, near Bath. Having only met once in over two years, we were engaged in developing designs for the 65-bed nursing home and 18 Extra Care apartments, which is a significant project for a small regional practice.

Working with the lead architect the nursing home plans were developed using wall styles and repeated libraries of parametric windows and doors, which could easily be replaced as the design developed through many iterations. The Helidon tool in the latest version of Vectorworks made it simple to investigate potential overshadowing issues on the communal gardens and also determine where external shading might be required.

On commercial projects of this scale accurate area schedules were also required. These were easily produced using the fully customisable Space tool in Vectorworks Architect. Door and Windows schedules were generated in the same way, with intelligent two-way data flows ensuring schedules could be amended and drawings automatically updated, saving time and eliminating errors.

Acoustic and structural issues were issues dealt with by other consultants who were provided with 2D DWG files and the BIM in IFC format. Industry Foundation Classes or IFC format is possible to import and export using Vectorworks Architect and allows the model's 3D geometry and data that define the building elements to be shared.

This opens up the possibility for energy analysis and clash detection, although at this point there are still problems with exchange between different software packages. Using IFC capable software such as Solibri Model Viewer, the project manager can integrate information from different consultants and it is envisaged that for future projects the model will be handed over at completion for use by the facilities manager. We can foresee this becoming more important as BIM becomes more widely adopted by end-users clients.

THE PROJECTS

The following cases explore how we used BIM on a small and medium-sized project and how early models were able to successfully communicate our design ideas to the client and the planners.

▶ PRIVATE HOUSE IN LEE BAY, NORTH DEVON

At the smaller end of the scale is our use of BIM on a residential replacement dwelling on a cliff top site in Lee Bay, North Devon for the design process and communication between the client and planners. Located on the beautiful South West Coast Path in an Area of Outstanding Natural Beauty, the project was very sensitive environmentally and contextually.

The existing single-storey bungalow on the site had severe structural issues, making it beyond economic repair, due to poor construction and some suspected ground movement.

Following a long battle with the insurance company, the owners sought to replace it with an environmentally sustainable dwelling, taking into account the scale, layout and appearance of their new home.

A survey was commissioned in 3D form and a 3D site model produced using the Vectorworks Architect Terrain modelling suite with the surveyor's data. This allowed the existing property to be accurately modelled and initial massing studies of proposed building forms to be produced – very useful for exploring early stage design options for discussion with the client and planners.

Once a basic strategy had been agreed, detailed proposals were developed with slabs, walls, parametric doors, windows and roofs, allowing plans, elevations, sections and 3D perspectives to be produced directly from the model.

Because of the sensitive nature of the site, the planning officer asked for a number of contextual views from key vantage points, resulting in some changes being made to a critical elevation.

Figure 2.8
Perspective views of the proposed building in context submitted for planning as additional information showing the building in context

The roof ridge height was also reduced to match the neighbouring property. Following a lengthy period of consultation, planning was granted and we were commissioned to undertake working and Building Regulations drawings, which were developed by adding more detail to the model.

Once the structural engineers had produced their designs for the ground beams, including 22 piles, these were modelled to incorporate into the BIM so that the sections accurately reflected the appropriate details.

At the time of writing the project is currently on-site and due to client changes, an application for a non-material amendment was required to introduce more glass and corner windows to the seaward-facing gable.

By updating the model, it was simple to explore the options and present these in 3D form to both the client and planner. It was possible to demonstrate that the changes would not be visible to any of the surrounding neighbours, so the planner was happy to approve the change quickly as a minor amendment at this late stage!

Another challenging aspect of the project was finalising the services strategy, as there was no mains gas available.

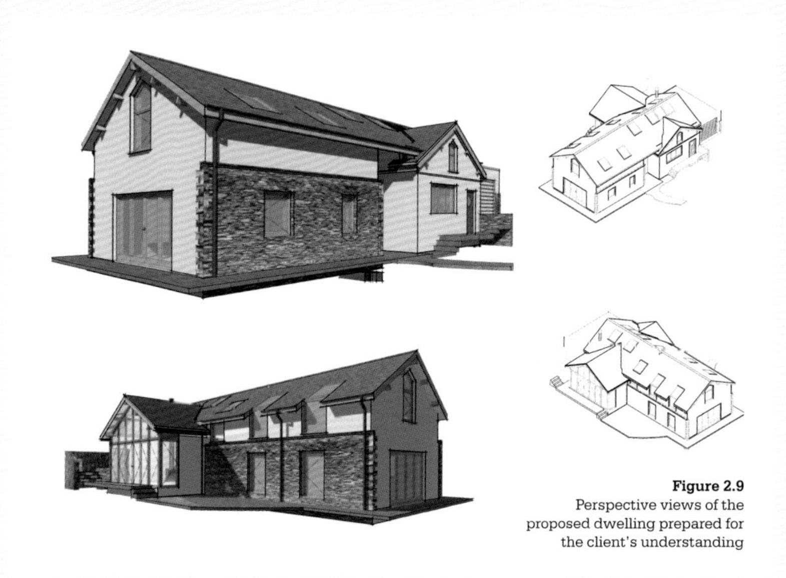

Figure 2.9
Perspective views of the proposed dwelling prepared for the client's understanding

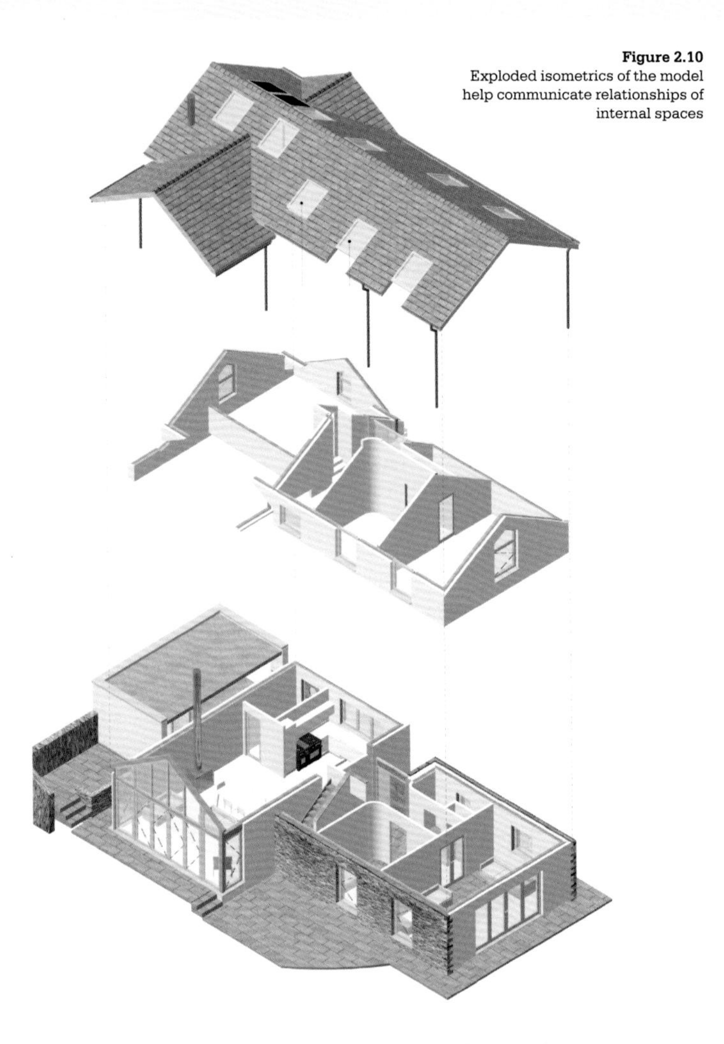

Figure 2.10
Exploded isometrics of the model help communicate relationships of internal spaces

Originally an air source heat pump was specified in combination with solar thermal panels and underfloor heating throughout. However the client became increasingly concerned about potential noise, having developed a hearing condition that made them very sensitive to background noise.

Ground source heat pumps using 65m deep boreholes were also explored, but this was ruled out by the consultant geologist. This left photovoltaic (PV) and the relatively new technology of thermodynamic panels as potential options.

Again the BIM was invaluable in assessing the visual impact of the panels and also using a virtual Helidon built into the software to analyse sun shadow patterns during the course of the year to demonstrate their potential effectiveness. All of the heat source options were also subject to SAP assessment by a consultant assessor (Synergy), who also made use of the information from the BIM, following the design changes.

Overall, using a BIM workflow on this project from inception right through to on-site construction certainly presented challenges, such as modelling some of the junctions between the roof and walls, using non- standard, bespoke windows and doors and also dealing with the complex site levels. However, it did allow late design changes to be incorporated relatively easily and communicated with the entire design team, including client, planners, contractors and structural and SAP consultants.

▶ THREE TERRACED HOUSES IN ILFRACOMBE, DEVON

Following a relocation from Bristol to coastal North Devon in 2009, we were keen to develop new relationships with local clients. However, while a number of new contacts were developed it was clear that North Devon fee structures were very competitive, especially when first trying to tempt a potential new client to move from their existing firm.

After being approached by a developer who already had planning for two traditional terrace houses, on a tight town centre setting in a conservation area of Ilfracombe, we were asked to look at the possibility of getting planning for three units. An initial concept design was required by the developer in a very short timescale and for a very low initial fee!

Once again the benefits of BIM were exploited to produce a convincing concept design that persuaded the client to commission a new planning application for the terrace of modern three-storey townhouses.

Key advantages of BIM workflows are the speed and efficiency of producing full sets of coordinated documentation and 3D views, in a way that traditional 2D drafting does not allow. This means it is possible to offer much more for less when the commercial pressure is on, making a practice more competitive.

Following planning approval the same model was developed to produce building regulations and working drawings, schedules and information on quantities, which helped the developer to look at costs of alternative forms of construction and assess these at a relatively early stage.

Detailed 3D models were produced of the bespoke elements such as balconies and lead roof dormers. This allowed very detailed material take-offs and some fabrication to be streamlined. For example it became very simple to model all of the eco floor and roof trusses required for the project, which resulted in detecting some clashes where waste pipes needed relocating, avoiding potentially expensive alterations on-site.

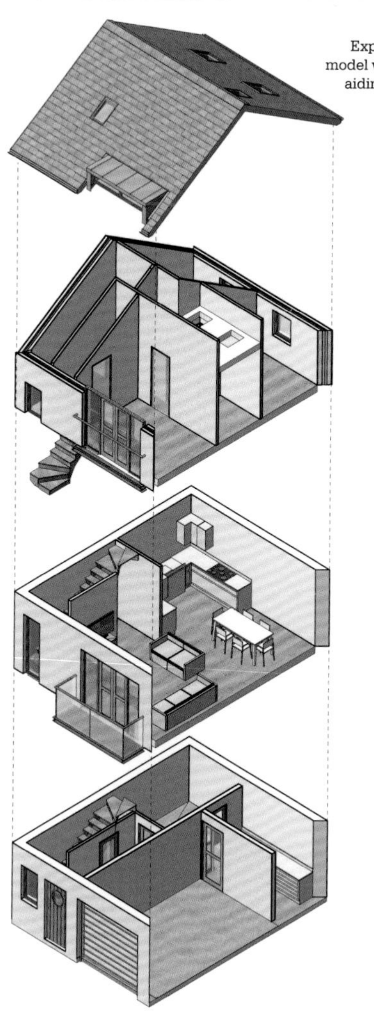

Figure 2.11
Exploded isometrics of the model which proved useful for aiding design development discussions

Accurate quantities of materials, trusses and also window and door schedules were produced from the model, making managing costs much easier too.

Another great benefit was the ability to start generating interest in the development by using the rendered perspective views of the scheme, prepared for planning, which made the estate agent's job of marketing the properties much easier which in turn pleased the client.

The buildings are now complete and make a welcome addition to the developing modern architectural movement emerging in Ilfracombe. More recently, following the positive regeneration occurring after the installation of Damien Hirst's 65 foot high 'Verity' statue, other interesting modern buildings such as a new Weatherspoon's pub have been approved, which bodes well for wider investment in this beautiful Victorian seaside town.

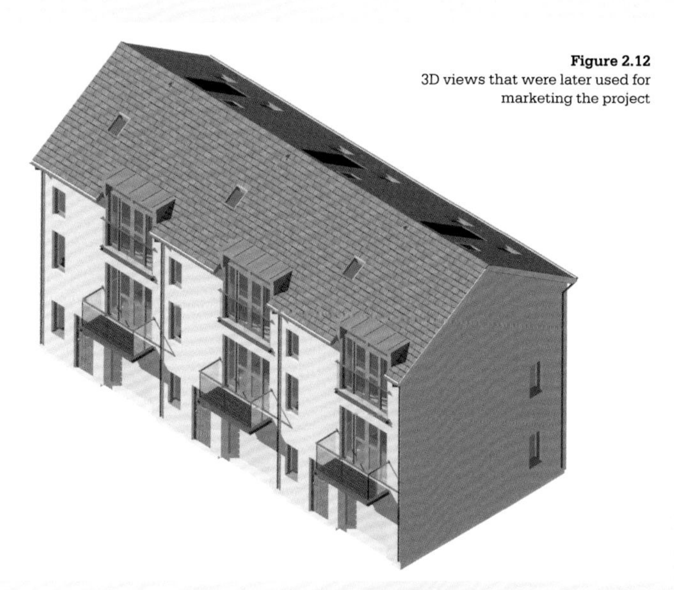

Figure 2.12
3D views that were later used for marketing the project

'BIM is all about setting standards and operating procedures that make collaboration much easier, simplifying the task of sharing data with other professionals who use different technologies, different file formats and different ideologies.'

CONCLUSIONS AND FUTURE PLANS

Having been a Vectorworks user since 1991, I have always been fascinated by the potential of 3D modelling, computer graphics and multimedia in architecture to develop and explain design ideas.

Certainly, the visual benefit of being able to develop, evaluate and communicate design ideas clearly via the 3D model using rendered visuals and animations has proved invaluable on all of these projects.

One might say that the use of BIM on these projects contributed no more than traditional workflows and 2D drawings would have. However that is to miss the point entirely. BIM is all about setting standards and operating procedures that make collaboration much easier, simplifying the task of sharing data with other professionals who use different technologies, different file formats and different ideologies.

All have simple aims in mind – saving time, money and effort – and improving the architecture that emerges, whilst also enjoying the process! In the near future, we are looking forward to seeing more of the industry adopt BIM workflows and hope to do more collaborative BIM with either other architects, or the consultants we work with.

We are interested in using our models more easily for early stage energy analysis with IES software (Integrated Environmental Solutions), who have recently agreed a development partnership with Nemetschek.

LESSONS LEARNT

- Don't make excuses for not trying BIM. Now is the right time to get started or keep developing elements of a BIM workflow and try to push the boundaries with every project.
- Investing in your practice's core CAD and BIM skills and the software is probably one of the most important investments you can make in your business.
- Before long, there is no going back to 2D drafting as 3D BIM workflows become far more productive, efficient, enjoyable and appreciated by your clients and consultants.

BIM AND CLIENTS

Timothy Ball (left) and Jon Hughes (right)

Susan Room

Practice:
jhd Architects

Location:
Cranbrook, Kent

Director:
Timothy Ball

Employees:
2 architects, 1 business development director and several freelancers

Founded:
2003

Technology:
ArchiCAD

03

There are many ways of making design options available to the client using the internet and this chapter focuses on using tools for sharing work online with services including BIMX and Dropbox.

It also offers a valuable insight into the process of outsourcing work. In addition there is a detailed look at what the practice produces at each design stage, demonstrating that BIM has a place in their deliverables at every stage.

'Imagine a pencil that could draw the construction of a wall, tell you how much of each material is in that wall, how well it performs thermally and give you every conceivable viewpoint of it.'

THE PRACTICE

In 2001 my business partner Susan Room and I started jhd Architects (a RIBA Chartered Practice) in collaboration with Sevenoaks architect Jon Hughes, of Jon Hughes Design. I am an architect and BIM specialist and I work with Jon closely on most of my projects.

I also have a Business Development Director and several freelance assistants, including ArchiCAD users, in the UK, India and Serbia. We are one of very few Kent practices using 3D BIM to design all projects. Our main area of work is high-end private residential, but we also take on other projects. For example, we are currently extending Hartley House Care Home in Cranbrook, Kent, to provide new communal areas and additional accommodation for residents suffering from dementia.

I have over 30 years' experience of running successful independent practices. I have been using ArchiCAD as a 3D design tool for 15 years and as a full BIM tool for the last four years. I've also had extensive experience working on BIM projects with contractors

Figure 3.1
New reception at Hartley House Care Home

to whom I'm now offering my services to help them take full advantage of the benefits of BIM. For example, a specialist fit-out company is currently using me as their BIM specialist to pitch for work with Lendlease, Stanhope and McAlpine.

The main driver for use of 3D design and BIM has been to improve the communication of the design to clients and contractors. The better we explain it, the fewer misunderstandings there are. As John Ruskin said: "Quality is never an accident; it is always the result of intelligent effort."

We started using CAD software in the mid-1980s, progressing from Turbo-CAD to AutoCAD and then to AutoCAD AEC, which was a fledgling 3D design package. However the architectural design tools seemed to be written as add-ons for what was essentially an engineering product so we looked for an alternative.

Now we use ArchiCAD, which is a full BIM design software package and we have an excellent relationship with the support staff in the UK, using their telephone support line several times a week. I invested considerable time and effort to learn the BIM aspects of the software as I think there is clear evidence that it drives productivity and competitive advantage.

Although we review our software options regularly, we have not found any better software on the market. ArchiCAD was written from scratch for architects and our tests prior to buying illustrated how much easier it was going to be for us to use.

Its user interface is very intuitive and visual. It is a very complete package in which we create 3D and 2D drawings as well as 3D renderings and 3D virtual building files all 'out of the box'. For instance, we find it easier to create a presentation document using ArchiCAD than using PowerPoint or Word. The only other software we use heavily is Excel, to export BIM data in text form for contractors to use. We also use Dropbox to issue PDF drawings and 3D files to clients, consultants and contractors.

Rather than creating super-realistic renderings, we prefer to spend time on the 3D virtual walk-through files, which our clients really like because it helps them to understand their building.

The 3D design aspect of the programme is very intuitive and in fact quite seductive in use. You genuinely feel that you are creating a building in 3D before your very eyes. However it takes a little time to get used to thinking in 3D geometry. You need to be very precise to make sure everything fits together so your geometry needs to be millimetre perfect.

When you draw using a 3D programme, you are effectively creating a model that the software then interrogates to create plans, elevations, section and other views from it. The link is live, so any changes are immediately reflected in every view. However you need to work out your own techniques for drawing graphics presentation.

The complexity and inter-relationships of the different graphics options at the start are quite daunting, but of course once you find a presentation you like, you can use it as a template.

The main external package we use is an add-on produced by CADImage (a New Zealand company). This tool creates a database of the project specification which is maintained within the project file. The text from this database can then be presented on different drawings and as traditional specification sheets, all in different ways depending on how much detail you need. So on a drawing you can show a short specification description and include the full description in the specification.

The same data can be exported as a text file or spreadsheet for the contractor to use. We don't use the NBS specification packages as we find them over-long and too complex for the types of building we design.

In summary, ArchiCAD is working very well for us. Creating a virtual landscape has become nearly as easy as sketching and far easier than physical model building. Imagine a pencil that could draw the construction of a wall, tell you how much of each material is in that wall, how well it performs thermally and give you every conceivable viewpoint of it. Whilst there is always a place for pencil and paper, from an efficiency standpoint there is no more efficient method than BIM.

The drawback is that you need to work on your 3D design skills in the same way that you once worked on your sketching and drawing skills. To get to the point of easy fluency in using the software probably takes a year or so. However the investment in time then pays off very quickly. Think of it as training a new assistant to work for you for a year.

The software also needs to be used by everyone working in the practice, from concept designers to technicians to job runners. It is only then that you gain the full advantage of the integrated building model that BIM offers. That requires changes in working practices.

In ArchiCAD you can have many people working simultaneously on the same model, from different geographical locations.

OUTSOURCING OUR BIM

ArchiCAD software is well used internationally, not least in countries where there are lots of freelancers. Its website has a job board for users to exchange services and some years ago we used it to source freelance help in India, Serbia and the UK. Hiring people on an ad hoc basis means we can control our workload, scaling up or down as required and managing our overheads very tightly.

To build trust and confidence, we have started by outsourcing the conversion of 2D survey information into 3D models. Our Indian and Serbian associates are very good at 3D modelling but they need a lot more guidance, especially when embedding the information to comply with our standards and working methods. We are helping them to improve the quality and accuracy of the embedded data – as this is becoming increasingly important – by introducing them to the concept of standard templates that incorporate embedded features such as our layer standards and IFC parameters.

Our UK associate is a working mum who I met at an ArchiCAD Summer School. She was starting to use ArchiCAD, but was already experienced in creating 2D working drawings. We are currently training her remotely to create BIM construction models from which we can extract the detailed drawings, schedules and quantities required by contractors.

'Rather than creating super-realistic renderings, we prefer to spend time on the 3D virtual walk-through files, which our clients really like because it helps them to understand their building.'

OUR STANDARD BIM OUTPUTS ACROSS A PROJECT LIFE CYCLE

We use the full 3D capabilities of BIM right from concept design stage, through detailed design and planning drawings to construction documentation. Key types of output for the various stages are:

- **Stages 2 and 3** Concept and Developed Design

- Simple 3D images created in much the same way as Sketch-Up but smarter to analyse multiple concepts against the brief.
- Site meshes based on land contours grabbed from OS maps.
- Massing elevations to compare, say, existing and proposed buildings.
- Automatic floor area and volume calculations using a tool called Zones.
- Presentation sheets that can combine drawn information, schedules, written text, PDFs and JPGs, all on the same sheet.

- **Stages 3 and 4** Developed and Technical Design

- Detailed design in 3D that really works – no faking it.
- Accurate plans, elevations and sections to illustrate the design, particularly to planners.
- Very quick 3D model output that clients can upload, on their iPad, Android tablet, laptop or PC and then view, play with and walk around their building in their own time, at home, or on the move.

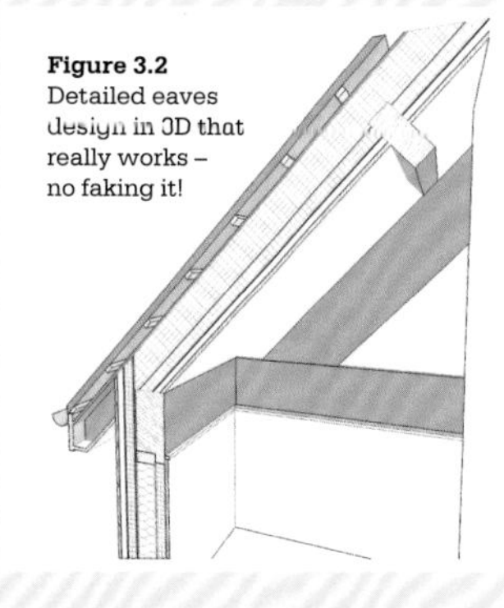

Figure 3.2 Detailed eaves design in 3D that really works – no faking it!

- **Stage 4** Technical Design: Construction Documentation

- Accurate and fully coordinated 2D plans, elevations and sections of the building.
- 3D and 2D details showing the construction. The 3D aspect is very powerful and also makes the explanation of junctions easier.
- Work packages created for any trade combining 2D, 3D and scheduled information. For instance we can issue an electrical layout that shows the locations of the power and lighting on a plan, 3D details of critical setting out areas plus a schedule of the number of power points and different light fitting types, including pictures, all embedded into the scheduling.

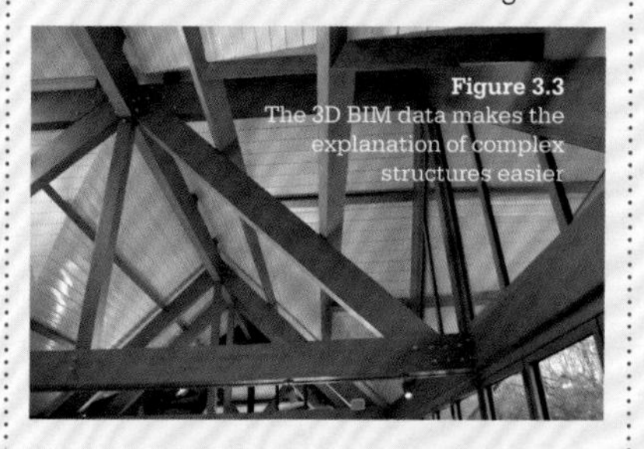

Figure 3.3 The 3D BIM data makes the explanation of complex structures easier

- **Stage 5** Construction: On-Site

- Accurate automated drawing registers to keep track of information issued and its status.
- Automated uploading of whole drawing sets to Dropbox. We can create a set of drawings for issue and automatically output them as PDFs without any need for printing.
- We are increasingly creating drawings that are optimised for viewing on laptops and tablets and very rarely use A1 size drawings. Most are now A4 or A3.
- The software also links drawing markers on one sheet to referred drawings. So you can have a multi -page PDF that cross-references from plan to section to detail to technical specification, all embedded in the PDF. We think that in the future we'll be able to improve this further using embedded html links.
- We also issue the 3D virtual building file to the contractor and they use their own laptops or tablets to check their understanding of the detailing on-site and illustrate the design intent to the site workforce. Incidentally there is no issue about accuracy because this file is created from the same information that creates all the other documentation.

- **Stage 6** Handover and Close-out: Record Drawings

- We update the 3D model throughout the project so that at the end of the job we can issue the client with up-to-date record drawings and 3D models for future reference.
- We are working on ways to embed user instructions and certifications within the same model. This will become easier when we can use html links.

THE PROJECT

The project is a contemporary building housing an indoor swimming pool, changing areas, double garage and a gym. It is located in the garden of our client's Grade 2 listed house in a small village in a Kent conservation area. The L-shaped plan wraps around the parking court, cut at an angle where an existing tree is located. Swimmers look out to the garden and in good weather two large sliding/folding doors open onto the terraces.

Our design was built using materials sympathetic to the conservation area. For example, the roof material is dark grey slate to blend with surrounding buildings and the external cladding is grey stained larch fixed with exposed stainless steel screws. The external windows and doors are grey powder-coated aluminium. The clay ground conditions and nearby trees meant that deep mass concrete foundations were required; however, above ground, the building structure is mainly timber. The scheme commenced on-site in November 2011 and was completed in August 2012. Completion was delayed by about four weeks due to groundworks problems with foundation depths and weather.

Figure 3.4
Two large sliding folding doors open onto the terraces

Figure 3.5
A BIM rendering of the contemporary pool house in the garden of the client's Grade 2 listed property in Kent

Figure 3.6
BIM meant that we could make changes quickly, knowing every drawing and detail was fully coordinated

We used BIM for the detailed design and planning stages and for the construction drawings and have made the minor changes needed to the model to reflect the as-built situation.

Although the client didn't request it, BIM benefitted their project in several ways. For example, an early meeting with the client and contractor led to a decision to relocate the pool cover housing to the shallow end and move the pool by 1m, thus increasing the space around it. BIM meant we could make the necessary changes very quickly, in the knowledge that every drawing and detail was fully coordinated and reissue the drawings whilst keeping the programme on track. The client was both reassured and impressed.

Figure 3.7
An external view as finished

BIM AND THE CLIENT

We are finding that more and more clients are attracted by the idea of having a 3D model on their iPad so they can be involved in the design at an early stage. Not all clients want that, but for those that do, BIM is a key differentiator which helps us win work.

Most clients generally understand the concept and value of designing in 3D, not least because it helps them visualise the building much better than traditional 2D drawings can. What they struggle to see is the value of the information embedded in the model and that of course is the 'I' in BIM. At Stage 3 we use BIMX (a model file type that can be viewed either on their computer or tablet to give them easy access to their 3D model).

But they can only see the design in 3D, not the data embedded in it. The only way they can see this data is either by using an IFC viewer, such as Solibri or BIMsight, or spreadsheets, each of which requires a better grasp of technology than some clients have.

We don't charge our clients extra for BIM, but it is reflected in our fees, which aren't the cheapest. We give clients BIM whether they like it or not because we know it helps them to understand the design and consider their options and reduces the number of misunderstandings on-site.

For example, we recently helped a great London charity, The Bromley-By-Bow Centre, to re-plan two areas of their building, using BIM to explain the different layout options.

We built our model without visiting site using photographs and 2D plans which they sent us. It was quick, accurate and extremely time efficient for all of us, not least the client, who immediately understood and reacted to the design. The model was used to explain various options for the layout of the space and the type of furniture that might work for their needs. It was reviewed by phone and the final updates issued immediately by email.

We are starting to present projects to planners using BIM but the technology they have available to them is very restricted. So, we've started to create videos of our planning submissions because they are accessible to everyone.

BIM also paid dividends at the planning stage. Firstly, we knew our design worked correctly and that we could build it, without having to go back and deal with changes resulting from the working drawing stages. This was particularly important as the design was very challenging in terms of the 3D geometry, particularly of the roof structure.

Secondly, the full colour elevations and 3D perspectives of the building we submitted helped sell the scheme to the Local Authority, as did our photo montage of how the building would look when viewed from the centre of the village. Generated from the model, these realistic external views helped the planners and the local community to visualise our intent and gained their support, which was particularly important given the pool's location in the garden of a listed building within a conservation area.

Figure 3.8
We used BIM on the project because it was challenging in terms of the three-dimensional geometry

After planning, we modelled the roof structure accurately in 3D, which allowed us to coordinate the design much more easily with the structural engineers. Although they were unable to work in 3D, the model helped them to think about the design in 3D.

We drew the structural design in draft and then updated the member sizes based upon their calculations. We did not detail the junctions and we really should have done. The junctions in some areas were very complex and the 2D structural sketches were difficult for the suppliers to understand. Next time we will do those as well.

Using BIM meant that once we had created the 2D elevations of all the roof truss details, we could generate schedules of the sizes and quantities of all the timbers needed at the touch of a button, saving time, money and potential delays and errors.

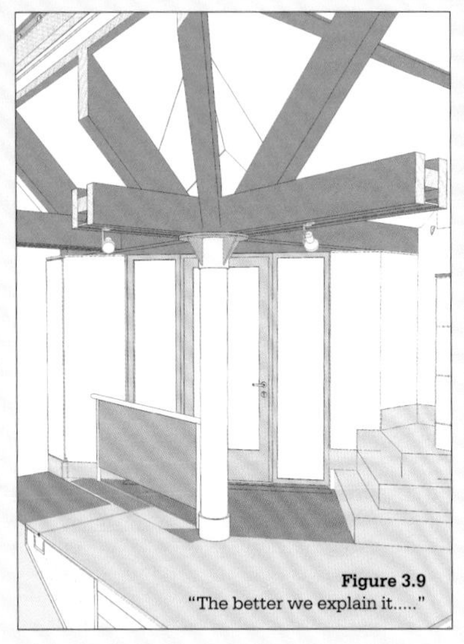

Figure 3.9
"The better we explain it....."

Figure 3.10
"...the fewer misunderstandings there are."

We also modelled the below-ground structures to ensure there were no clashes between foundations, drainage and the ducting both for the air handling system and the pool water treatment system.

The use of BIM also helped with the design and location of a sculptural pendant light fitting. We were able to import the 3DS file created by the designers directly into our model so the client could visualise the fitting and we could work out the correct location and height to fit into the space.

In the early days, this level of detail meant that the time we spent at working drawing stage occasionally went over budget, but the time on-site reduced so, as long as the project got built, the overall fee time remained under control. We didn't try to recover these costs, as we viewed the time spent learning how to create the models as a long-term investment in improving productivity. Now that we are good at it, using BIM takes no longer than conventional drawing and our aspiration is to reduce the time spent as we get better at it.

Figure 3.11
BIM helped us with the design and location of a sculptural pendant light fitting

The fact that our chosen contractor had not used BIM before was a potential challenge, but their enthusiasm and willingness to work collaboratively meant that they learnt quickly. They used Dropbox to access all the latest versions of the drawings and had the 3D model on an iPad on-site. This meant they could really understand the design intent and detail of every aspect of the building, including difficult junctions. As a result, they delivered the project to a high-quality and there were fewer potential costly errors and misunderstandings.

To sum up, using BIM on this project really helped the client, planners and the contractor fully understand the building. We learned the value of getting the model right and using it to create 3D details which we had not done before. Everyone on the project was complimentary about this way of working, including the Building Control Officer who put the building forward for an LABC award.

CONCLUSIONS AND FUTURE PLANS

BIM means we can take on more complex and larger projects without having to expand our staff. It has allowed us confidently to accept more work because it is equivalent to a full-time assistant. Because of the advantages of easier and more accurate coordination, we have not increased our fees. In a sense the BIM investment pays for itself. It allows us to offer it as a key point of difference to both clients and contractors.

However it can be a challenge persuading clients to buy into the technological approach as many still feel that they struggle with email!

It also affects the way we work in terms of workflow. It puts pressure on the earlier stages of the project for accuracy both dimensionally and in selection of construction techniques. As a result we may need to create, for instance, a wall type at the early design stage, based upon likely construction and thickness, and then update the wall type as the design progresses.

We have tended to work with the software rather than customise it. The exception is the creation of bespoke template files that structure the complex array of 2D and 3D views as well as the building materials we commonly use. The templates 'out of the box' are very generic and need thought to facilitate efficient creation of the type of buildings we design.

We are also now in a position to help contractors working collaboratively on design and build projects to create models of their work packages. The BIM can then be exchanged with a main contractor who is responsible for maintaining the full model of the building. This not only gives us additional work, but also introduces us to larger contractors and new business opportunities via them.

Unfortunately we have not found other consultants such as structural engineers, QS or M&E taking up the 3D model idea sufficiently, so we have to do some of their work for them. Our structural engineers routinely now give us sketch plans of the structural layouts only and then we create the structure in our 3D model. We are then certain that it all fits and can also schedule the structural members for easy costing.

This additional work can, however, be reflected in the fee allocation – by charging the client more for effectively doing some of the structural engineer's work, but with a reduction in the structural fee so that the overall cost remains the same. The structural engineers, however, remain responsible for checking our structural layouts and schedules for accuracy against their design. The real benefit to the architect of doing this is the full integration of the structure into the building model and the consequential improvement in accuracy and reduction in site problems that result. You may do more work, but it saves you time and aggravation later.

In the future we will hopefully find more engineers who work in 3D and we can then exchange 3D information with them. Most of the steelwork suppliers are already creating their fabrication drawings in 3D so that would 'join up the dots'. I'm sure other engineers will follow suit.

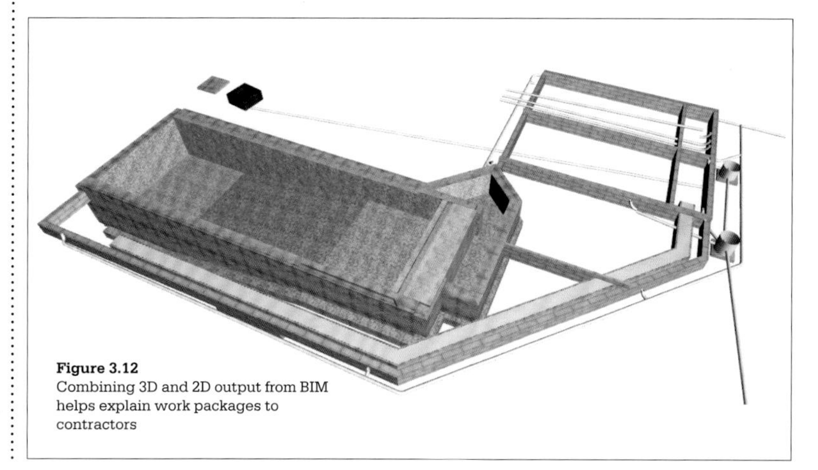

Figure 3.12
Combining 3D and 2D output from BIM helps explain work packages to contractors

LESSONS LEARNT

- Plan a year-long programme of getting to know the software. That can be challenging financially, but will pay dividends after that initial time investment.
- Use online video training rather than face to face. It's cheaper (some is actually free!) and just as effective. It can also be accessed when you have the time.
- Stay involved with the software community and take part in discussions to keep learning and challenging the way you do things – don't be afraid of asking simple questions.

TRADITIONAL AND MODERN METHODS OF CONSTRUCTION USING BIM

Rob Annable

Practice: Axis Design Architects Ltd

Location: Birmingham

Employees: 2 directors, 1 qualified architect, 1 Part 2 architectural assistant and 1 admin support role

Founded: 1984

Directors: Rob Annable and Mike Menzies

Technology: Graphisoft's ArchiCAD and Tekla BIMsight

04

The case studies described here explore two different types of housing project of different scales and complexity. Each sought to capitalise on the opportunity for sharing more informative construction information with other design team members and each brought challenges regarding file type, presentation, drawing and data format.

The first is a design and build, public sector housing project on behalf of a local authority developed by the team at Axis Design; the second is a personal self-build project by Rob Annable. They have been chosen to demonstrate how BIM can improve both traditional and modern methods of construction and be an asset to varying scales of practice.

THE PRACTICE

We are based in Birmingham city centre and have recently completed our 30th year in practice. Formed originally by four lecturers and teachers who met during their time at Birmingham School of Architecture, the practice is now run by myself, Rob Annable, and the founding director Mike Menzies.

Our client base primarily consists of local authorities and housing associations and we have a reputation in the affordable housing sector for our design-led projects around the West Midlands, as well as our public consultation experience and dedication to sustainable design development. Exploring ways to use digital technology to improve our service both publicly and privately is also an important aspect of our work.

During the last decade we have developed numerous techniques for communication and information sharing in the built environment with our clients and we see BIM as a natural part of that business ethos.

We use Graphisoft's ArchiCAD for our BIM production and it has now become the primary tool for all drawing output in the office, across all stages of work. Our journey to full BIM adoption was a slow one, thanks to familiar concerns regarding cost and training causing us to hesitate. However we have now made the transition and are reaping the benefits of production improvements in both quality and quantity. With the benefit of hindsight we now wish we'd made the leap years earlier!

Access to BIM was available and encouraged at Birmingham School of Architecture in the mid-1990s (some time before the acronym was invented) thanks to a particularly foresighted tutor. The experience I gained during my time there as an undergraduate led to the practice experimenting with basic visualisation techniques and the early adoption of a single licence of ArchiCAD in the office at a very early stage compared to much of the UK market. Implementation was sporadic and simplistic, however, and much of the full potential of the tool went unused for the first few years, with the mix of software and protocols in the office at times becoming a hindrance.

The practice finally chose to commit to office-wide adoption and genuine use of BIM in 2010, deciding that an almost overnight implementation and intensive training strategy was the best way forward. This involved joint training sessions with all the staff attending a four-day course at the vendor's office and additional work using online tutorials and guidebooks in the office.

We were able to spread investment costs over a 2-year period thanks to the purchase choices made available by Graphisoft and no additional machines were required in the first year thanks to the robust performance provided by the software even on mid-range performing machines – both important factors in choosing to continue with ArchiCAD, alongside the director's ability to lead from the front and train in-house with a familiar tool.

A modest upgrade to RAM and video cards in existing machines was later required following the commission for a large, neighbourhood-wide model for public consultation. The latest version of ArchiCAD (17) now requires 64-bit machines, however, and this will result in the need to invest in new workstations in the coming year. Whilst the first six months of use were predictably challenging, with many lessons learnt regarding workflow, file protocols and output quality, it was also easy to recognise that the capacity of the office was quickly improving.

Better coordination between drawings and greater confidence in the accuracy of the information allowed all members of staff to undertake larger quantities of work and responsibility, resulting in more cost-effective labour. Investing time agreeing drawing formats, pen weights, schedule styles and data protocols proved crucial and took several iterations before file templates became fully successful. Collaboration within the office was also improved thanks to the use of joint working tools available in ArchiCAD such as 'Teamwork Server', with staff working simultaneously on the same file both within the office and, when necessary, remotely from home.

Teamwork uses a change tracking database on an easy to install central server that allows portions of a model to be locked or released to particular members of the team, preventing the danger of conflicts. Ownership of areas of the BIM content can be exchanged quickly between team members using a request and approval system across all aspects of the software, from model to schedules and drawing layout design.

Initially the benefits of our investment in BIM were predominantly felt internally, with modest changes to our clients' experience. As our confidence grew we were able to provide new types of information and experiences as a standard part of our service and this in turn assisted us in winning more work and sustaining our ability to provide a service in all work stages against competition from larger offices with more staff.

This, in turn, has allowed us to push the boundaries of our BIM usage and find new ways for the whole construction team to benefit. For the client this benefit begins with the increased clarity of information and improved dialogue, not only between them and the designer but also between the project team and the public. For example, using the same workflow to deliver public consultation information managed and developed by the same team responsible for planning applications and construction is of great financial and political value. For only modest additional fees we are able to provide live visualisations that can be controlled and interrogated by the public and promise their future delivery with confidence.

THE PROJECTS

▶ BIRMINGHAM MUNICIPAL HOUSING TRUST

- **Design development: planning applications and consultation**

Axis Design are one of several practices responsible for designing Birmingham City Council's new-build affordable housing projects. Our initial role requires us to submit a planning application on behalf of the City Council following a cross-departmental collaboration with various stakeholders in the local authority. We used BIM to prepare the application, drawing from a pool of house types that have been developed with the client over the last 5 years. They are referenced, or 'hotlinked' (the ArchiCAD term for a linked external reference to another live file) into a larger site model allowing us to model in great detail within the individual house design whilst simultaneously developing the site strategy. This also allows us the opportunity to benefit from other associated BIM techniques such as the importing of survey data text files to automatically generate a 3D mesh to accurately model the existing ground level.

During the design development meetings with the client we have used BIM as a live design tool to help with decision making on particularly complex sites, demonstrating immediately the impacts of the proposal on the surrounding neighbourhood. On larger projects we have also provided a model of the proposal for use in public consultation exercises, moving beyond the traditional artist's illustration towards a more democratic, informative depiction of the scheme using a live 3D model that can be viewed freely in the same fashion as first person gaming technology.

This is possible thanks to a complimentary piece of software available from Graphisoft called BIMx. The ArchiCAD file can be exported in a fixed format for anyone to view on their own computer without the need for any investment in additional software. During an afternoon consultation event, for example, we provided access to the public and clients on laptops and projectors in a community centre and the simplicity of the interface meant that anyone could view the proposed buildings quickly and easily on their own terms.

The latest version of this tool is designed for use on hand-held tablets and available to the public as a smartphone 'app'.

The critical benefit here for both consultant and client is the ability to efficiently provide all these outputs in a single workflow, thus avoiding additional costs and time for duplication in alternative software tools or external skills required from other organisations. Whilst some additional work must be undertaken to model the development to a degree of detail appropriate for public use, the core drawing activity is simultaneously delivering planning application drawing information in traditional plan, section and illustration form. Furthermore, when undertaking the initial building design we embed information into the model regarding likely construction principles and this begins to form the basis of our work for the later construction information stages.

The example offered here is our recently completed scheme for flats and houses called Meon Grove, situated in the south of Birmingham. Comprising 19 houses and 12 flats, we completed the design development process with the client team during the early stages of our BIM adoption process. Site constraints such as protected trees and services put the hoped-for housing density under pressure and creative accommodation typologies were required to meet the brief in terms of both economic viability and space standards.

Our response included a three-storey block of flats positioned over the area of the site with the greatest level change, allowing us to raise the living accommodation over an undercroft parking area. Achieving this resulted in some complex geometry to coordinate through the circulation spaces and testing the proposal in BIM gave us full confidence in the proposal. Depicting this to the client with an accurate 3D model also ensured there was support from all stakeholders and the planning approval was secured with no concerns or objections raised.

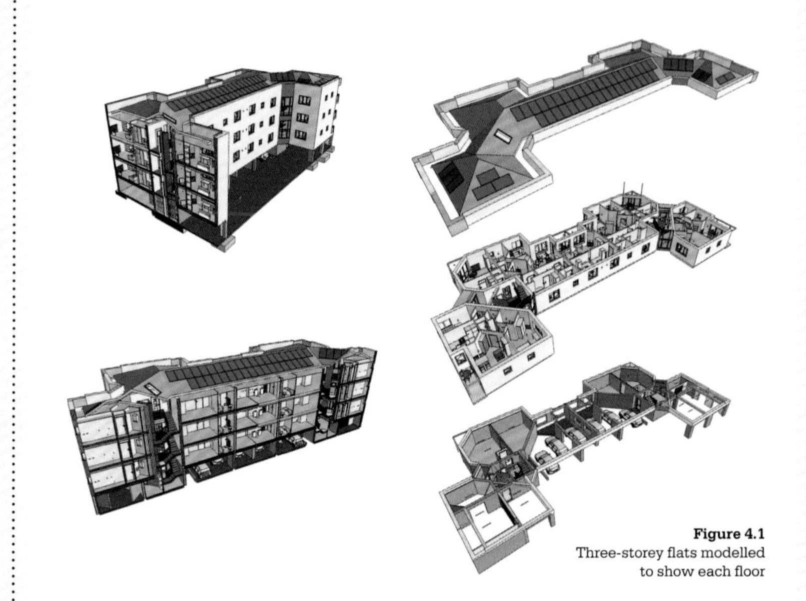

Figure 4.1
Three-storey flats modelled to show each floor

The approved planning drawings were then used by the client to inform a design and build tender with local contractors, again proving the value in testing the building layout in three dimensions at the earliest stage and reducing the risk for tenderers.

- **Construction information including structures and services**

We were appointed to complete the construction information for the successful contractor.

Funding deadlines driven by public sector spending targets created pressure to commence on site as quickly as possible and we were able to assist with a short timescale for initial information thanks to the degree of information embedded in our initial BIM package prepared for planning application stage. Standard house type drawings were provided swiftly and efficiently, allowing us to focus on the more complicated flats.

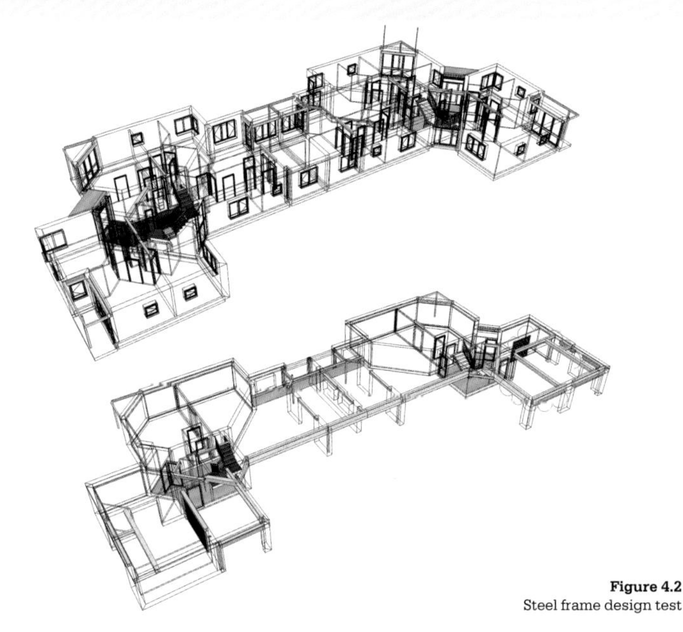

Figure 4.2
Steel frame design test

The BIM output benefits for both design team and client were numerous on this part of the scheme, ranging from complex spatial tests to schedule outputs that automated the counting of building elements.

Following input from the engineer the contractor decided to alter the intended construction method from masonry to steel frame. This necessitated intensive collaboration over the design of the steel section sizes to avoid clashes between beams and intrusion into designated service zones. To assist the engineer and the client we provided additional modelling of the steel frame, creating a visual model of all elements and a written schedule of steel quantities.

The model was viewed during meetings and also provided in IFC format and BIMx to the engineer to compare with their own 2D information. Annotated versions of the model were also provided using Tekla BIMsight, allowing them to make full use of the visual information and embedded data in each steel element. Whilst final responsibility for construction information remained with the engineer, the use of the BIM information during design prevented several instances of clashes between parts of the building fabric and helped rationalise the number of different size sections being used in the building.

A similar method was applied to mechanical and electrical coordination. Axis Design provided additional information to the design team thanks to the use of a dedicated services design tool for ArchiCAD called MEP. This allowed us to detail pipe and cable runs throughout the model, including connections and appliances. Although no scheduling information was utilised in this instance – with drawings provided being effectively schematic for discussion – we were able to provide informative, richly detailed drawings and also use a live demonstration of the model during meetings again. Unfortunately, unlike the structural information, no additional file sharing of IFC data was possible in this instance as the M&E subcontractor was unable to utilise digital information.

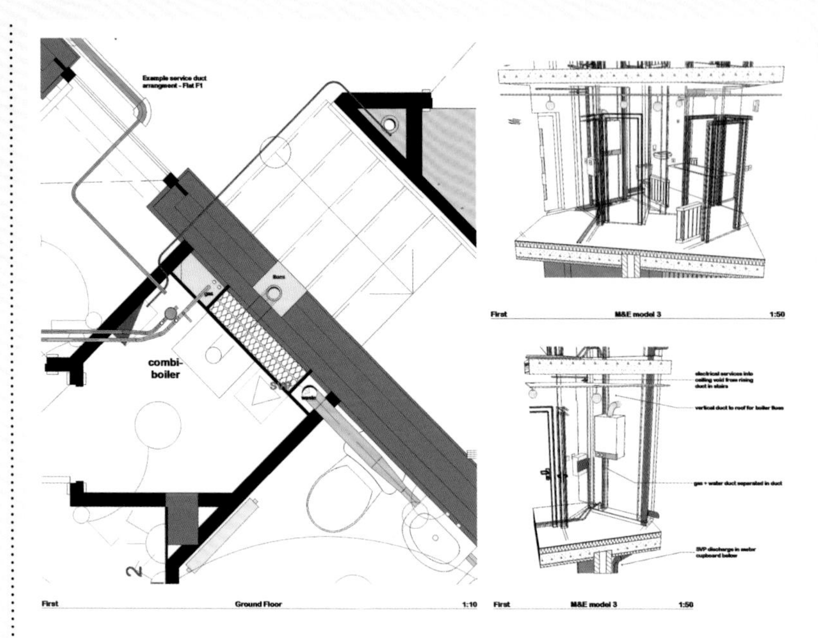

Figure 4.3
Services schematics tested in model

We are firm believers that BIM adoption should be seen as the natural progression of a traditional architect's role and offer its use as part of our standard services. The profession should offer clients increased value by improving service levels rather than cutting fees. However early adopters will likely find themselves negotiating complicated overlaps with other professionals. In this example the levels of information provided on structural and M&E information were offered at no additional fee in order to improve the communication for all involved and test our BIM capability. We hope that future projects will see greater input from other consultants so that time and responsibility for this type of information is fair.

Similar restrictions were experienced on site, with minimal access to a laptop for viewing digital information. However it should be noted that the value of informative 3D drawings on paper should not be underestimated. The site manager commented on how useful they were to assist in orientating himself through complicated areas of the building.

In summary, despite being one of our earliest challenges with the office-wide BIM strategy, and the first example of use right through to construction stages, we were able to generate a level of BIM detail that radically altered the way key parts of the building were discussed and designed. This avoided errors on site, simplified areas of the construction and assisted the entire team in interpreting the problems and challenges on the project clearly to the client, allowing them to make well-informed decisions.

▶ A SELF-BUILD PROJECT

As a demonstration of what is possible for the micro practice or sole trader when using BIM, our second case study is a self-build home for my own family. The experiences described here are all examples of drawing outputs and design interrogations undertaken single-handedly during weekends and evenings, alongside the larger-scale projects described previously.

One of the many agendas for the project is the exploration and experimentation in new building techniques that could inform future projects for my clients, using the house as research and development for the practice. The scheme tests off-site timber frame manufacturing against on-site masonry techniques, creating alternative methods of construction and finished environment. The final design of the house was arrived at after many iterations that were all tested extensively using ArchiCAD's numerous tools. Two key issues are common to most of the experiments undertaken: energy performance and manufacturing processes.

- **Energy assessment: Whole house performance**

The overall performance of the building envelope was tested with the use of Graphisoft's EcoDesigner tool. By using local climate data and entering information regarding material performances, use patterns and energy supply, a whole house assessment was completed in the early stages whilst the final specification was still undecided. This was used to set benchmarks against which minor alterations to aspects such as window locations and size could be tested. For example a range of comparisons demonstrating the impact on net losses or gains of solar energy was quickly tested as the design evolved. In later stages this was also used to provide a head start to the detailed calculations by exporting the building envelope data into other tools such as the Passive House Planning Package (PHPP).

The latest version of this software – EcoDesigner STAR – has just been launched at the time of writing, but we were able to take part in early beta tests of the product, which also allowed building details to be tested for thermal bridge problems directly from the model. This is a service usually requiring specialist input and additional software. Although the tool is an additional cost beyond the basic licence, EcoDesigner is embedded in the ArchiCAD interface meaning no other third party software is required and any issues of interoperability are avoided.

- **Manufacturing: Fabrication testing**

Having confidence in this energy prediction requires confidence in the construction quality, and the use of BIM has also ensured that the fabrication process of this build, both on and off-site, has been carried out as planned.

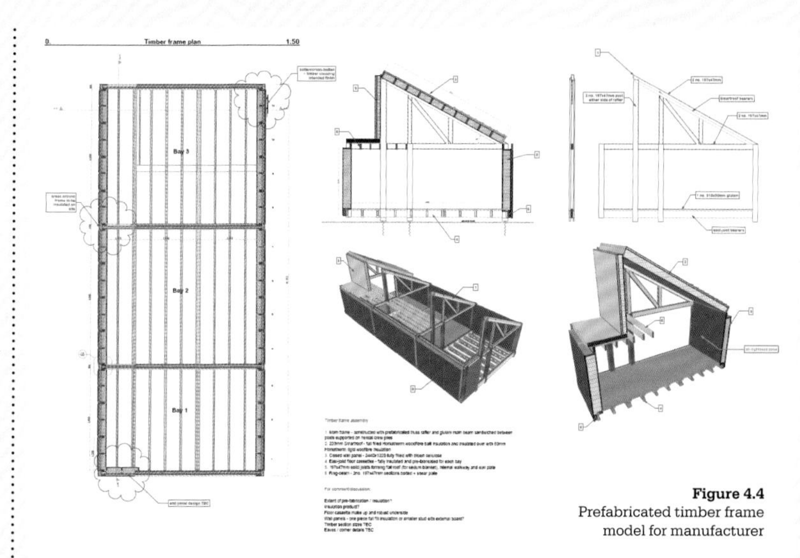

Figure 4.4
Prefabricated timber frame model for manufacturer

By collaborating with a timber frame construction company and an engineer who were both enthusiastic about the outputs from BIM, we were able to gain the maximum benefit from the time invested in a fully detailed model. With greater support from the whole team came improved communication and sharing of responsibility. Time was used more effectively thanks to discussion over a single drawing format and the successful use of online tools reduced the need for meetings or travel.

The timber frame element of the house was modelled in collaboration with the manufacturer to test for complicated junctions that might hinder progress on site and undermine the airtightness strategy. The same information was simultaneously used to provide the structural engineer with IFC information that could be imported into their software.

Once adapted this was then placed back into the architect's model for the final iterations before manufacture. Third party software solutions such as Tekla BIMsight were also used to provide the communication channel

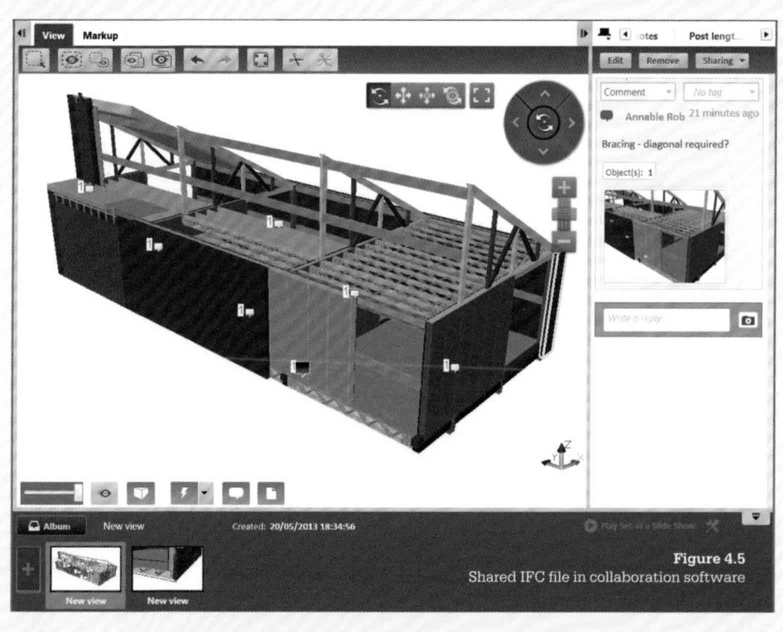

Figure 4.5
Shared IFC file in collaboration software

between all parties, avoiding any issue of interoperability. By importing the BIM data in IFC format into the Tekla software consultants were able to communicate with each other in a separate platform that allowed them to continue with their own chosen workflow/software alongside the commented file, which was updated and shared via the project management web site – basecamp.com. The finished product has recently been erected and the construction process was completed cost-effectively in under three weeks, demonstrating the value of full design testing in the model.

Figure 4.6
Prefabricated timber frame wall panel being lowered by crane

Whilst the drawing and modelling process for this project was similar to the techniques utilised in the previous case study, this project demonstrated during construction the benefits of combining BIM with MMC.

The increased accuracy of the production information was subsequently maintained in manufacture under more controlled conditions and only one minor error occurred, resulting in a ventilation duct being misaligned by approximately 20mm after a late decision to vary a beam size wasn't included in the model.

In addition to the manufacturing and energy performance benefits, the BIM strategy has also allowed a complex project to be successfully project managed thanks to the ability to measure material quantities and control procurement. The automated production of schedules ranging from traditional information such as window and door data, through to joist lengths and primary construction material volumes/areas, has successfully brought all the design decisions together to create the data needed to construct the house as fully intended in the earliest sketches.

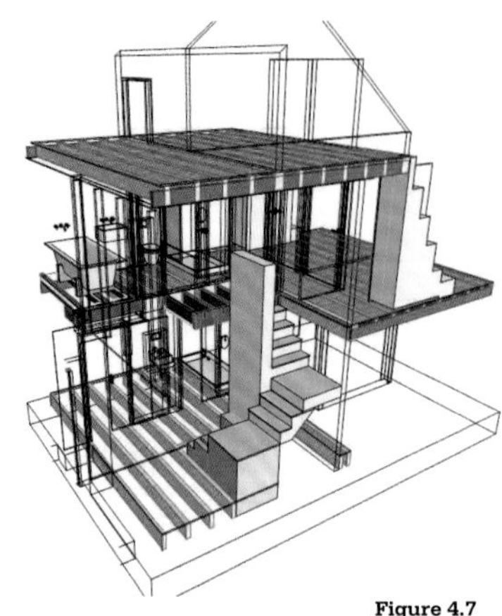

Figure 4.7
Joist schedule output from model

The ability of the architect to create this information within the same workflow as the tool which helps visualise the aesthetic of the building – examining poetic qualities as well as practical quantities – provides an opportunity to create more complex, better performing architecture thanks to more informed decision making.

'In difficult economic times for the profession we have provided greater value by raising our game rather than cutting our fees.'

CONCLUSIONS AND FUTURE PLANS

Although the initial capital cost of BIM software, hardware and training had to be carefully planned for, we have realised the benefit of that investment quickly by being better equipped to win more work and improve our productivity without additional investment in further staff or hardware.

Since adopting BIM within our office we have been able to improve the service we provide to clients across all aspects of our work. In difficult economic times for the profession we have provided greater value by raising our game rather than cutting our fees.

Value created from better service begets value across the whole discipline.

The comparisons offered here demonstrate the benefits that BIM can offer the architect working on both traditional construction and prefabricated MMC techniques. Similar methods of modelling and information were utilised but at different times of the contract, indicating that the approach to service levels, fees and resources should be considered carefully. A greater investment in extensive pre-construction testing allowed the full potential of the off-site method to be realised, consequently drawing other consultants and manufacturers into the process much earlier as well.

In the case of the traditional construction project the BIM exercise and outputs were more responsive to particular situations arising on site, resulting in a higher resources demand during the mobilisation stages. Although the service provided by the architect ultimately resulted in comparable outcomes, the difference in collaboration and dialogue with other team members was notable and shared ownership of the project was at times lacking. It's important to realise that the most valuable outcome of a BIM process is the communication it fosters, rather than the model itself.

The benefits of combining BIM production skills with MMC are clear. However our ability to deliver this degree of coordination is often hampered in the UK by the habit of split procurement strategies before and after planning consent. It is not unusual for a consultant team in the housing sector to be commissioned for initial design stages but then have no guarantee of also completing the construction information stages. Without this the levels of coordination described here are impossible to deliver and MMC less likely.

BIM can be the glue that binds a team together from beginning to end. We should promote its use to support better communication and continuous, longer-term responsibility.

LESSONS LEARNT

Key lessons learnt are primarily around the area of internal office working methods:

- Whilst large practices may be able to justify a full-time formal BIM management role, we would recommend at least the nomination of a BIM champion responsible for directing and shaping office policy on documentation and model construction.
- Although we successfully set up basic templates both within the software and our file server, with hindsight we would have benefitted from a more regular review and adjustment as we learnt from each project. With many aspects of drawing production able to be automated, capitalising fully on this function should be taken seriously.
- With regard to drawing techniques themselves, we have found that one of the most difficult issues to decide upon is when to stop. With every project there is a point at which the continued modelling of some of the finer details in 3D can in fact hinder progress. Deciding when to freeze the BIM and complete remaining details through annotation and/or more traditional 2D techniques is an important skill.
- Finally, our advice would be to press every button. The number of functions and tools available in most BIM packages can be daunting, but until you've tried everything you won't be sure you're getting the full benefit from the technology you've invested in.

WINNING RESEARCH FUNDING WITH BIM

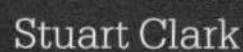
Stuart Clark

Jon Moorhouse

Practice: Constructive Thinking Studio Ltd

Location: Liverpool

Employees: 2 architects, 1 managing director and 1 architectural assistant

Founded: 2002

Directors: Jon Moorhouse, Hannah Moorhouse, Stuart Clark

Technology: ArchiCAD; BIM Server; BIMx; Eco Designer; Cigraph Architools

05

On this project, BIM was used both for coordination and as a design and assessment tool, to form the basis for a funding bid from the Technology Strategy Board.

This UK public body operates at arm's length from the government, reporting to the Department for Business, Innovation and Skills and is responsible for awarding grants to test, implement and develop technological solutions as directed by central government.

For this funding the Technology Strategy Board was looking to test solutions for retrofitting the country's vast existing housing stock with different technological measures to improve their energy performance by a significant 60%. The programme was called Retrofit for the Future.

THE PRACTICE

Constructive Thinking Studio was formed in Liverpool in 2003 by Jon and Hannah Moorhouse. Jon had used and taught ArchiCAD since 1992, as a sole practitioner. Stuart joined the practice in 2006. With both having used ArchiCAD since the 1990s it was only natural to continue using it within the practice. At the time ArchiCAD was one of the few stable software solutions facilitating the Virtual Building or BIM.

Our work is currently centred on housing (bespoke and social), leisure and some commercial, with an equal mix of refurbishment and new-build. Key current areas for BIM collaboration are within the social housing sector and individual housing designs with complex or bespoke construction elements.

Given that we were both early advocates of a BIM workflow and that ArchiCAD facilitated collaboration and teamwork from relatively early on, we were keen to develop our skills beyond normal practice expectations.

Training was informal – usually through trial and error – or on an ad hoc basis. We regularly work with user groups and attend annual conferences and workshops to extend our abilities and showcase our techniques. Together we have developed workarounds for our requirements and over the years, through feedback to the software developer, these have become new features in subsequent versions.

As the practice has grown and the hardware and software have improved, we have yet to find a competing software solution good enough for us to consider moving from ArchiCAD, now in its 17th iteration. (We have used this software since Version 3 so may have a slight positive bias). We work towards interoperability wherever possible, with all our models ready to be published as IFC with minimal additional tweaking.

From when we first started working with computers, we saw their ability to facilitate automation as being a great asset, beyond architecture and construction.

As architects we model, then build from our models and we saw adopting BIM as the best way to model better: build the model correctly and the information taken from it will be correct and coordinated, it's as simple as that.

Over the years things have become more sophisticated and the information more complicated, but this core principle remains the same. We were always aware that the limitations of software could create limits to our architecture and have therefore sought ways of modelling exactly what we want, through third party add-ons and our own programming where necessary.

We use BIM across all of our projects – from house extensions through to 80-bedroom hotels, on both new and existing buildings – using a range of survey techniques, depending upon the project size and complexity. On micro projects, this remains profitable through standardisation of our template and an almost automated output – we can fully model detail and schedule a bespoke single-storey extension in a few hours and amend it rapidly from there.

On medium-sized projects the benefits from modelling everything first are myriad, from accurate take-off to resolving the awkward part of that detail you just can't see and very efficient amendment processes. On large projects, coordination of information and server-based collaboration are key to the efficient delivery of our projects. We can also provide (limitless) additional information in terms of sections and details (to suit the needs of the contractor) very quickly as it's just a matter of adding annotation and dimensions.

The project discussed below will deal with how using BIM has enabled us to win funding and more generally how coordination using BIM aids our practice in our everyday work.

BIM SUPPORTS THE REALITIES OF OUR WORK

The architect's role as a designer is the one most people recognise. It's cool, it's sexy and it's easy to understand. It also glosses over the mundane day-to-day reality of architectural practice, which largely consists of preparing, collating and coordinating all of the information required to procure a building accurately, on time and within budget.

Traditionally, this coordination would be through the use of drawings (printed or digital) and a keen set of eyes. However the process is laborious and time-consuming and prone to inconsistency, particularly during updates, often where sets of plans/ elevations/sections from other consultants don't match. This does not happen if you're using BIM correctly. They're always consistent by default!

With all the talk about 2D, 3D, 4D, 5D... and COBie drops, the fundamental advantage of BIM seems to get lost in the noise. The real benefit is from the coordination method used to ensure the information is consistent. This means less time, energy and materials are used and misused on-site. Benefits such as visualisation, energy modelling and facilities management are in addition to this.

To begin with we find that a coordinated model allows us to spend more time on design, allowing us to assess, document and present a wider range of options in a shorter period of time, giving us a robust information set for persuading clients and stakeholders of the benefits of the final solution. It's like presenting your working out as part of an answer.

When a client asks why we have proposed a certain solution it's always helpful to be able to show them. Using BIM we only have to model each option once to have a full range of information, rather than drawing our plans, sections, elevations or sketches for each one. The benefit of the combination of software we use is that not only can we do this for design, but also for structure, programme, cost and energy performance options.

COORDINATING OUR INFORMATION WITH OTHER CONSULTANTS

Coordination always begins with our own information, in terms of traditional drawing issues, but as the projects progress, we incorporate that of suppliers and other consultants as well. Sometimes this is done directly into ArchiCAD though IFC but more often, it will involve some level of additional work on our part. Additional information from other consultants is always segregated, whether modelled by them or not, in order to avoid confusion. This coordination has always been a key part of the architect's role. Over the years, the way we do this has evolved from overlaying third party drawings and modelling their contributions if necessary, to the point where it is now possible, through IFC, to reference models into ours and coordinate our information in a much more direct manner, as well as track changes between the different versions.

This feature within ArchiCAD is great for picking up alterations which may have been missed from a revision note and is part of our internal checking mechanism when issuing models to third parties as either WIP or published information.

In reality, however, things are rarely that simple. For years we have been working in an environment where we are the only consultants working in this way, but this situation is improving. Whenever we work with other consultants and suppliers, we ask whether they work with BIM and request an IFC as well as the drawings. Normally the answer is 'what's an IFC?', or that their software doesn't support it, but adoption is becoming more widespread.

There are other formats that can be used to transfer 3D geometry such as .gbxml, .kzml, .dwg, .3ds etc., which are usually good enough for coordination, so we'll often obtain that as well as the drawings. We generally offer to help our partners with BIM.

COORDINATION BREEDS ACCURACY; BREEDS CONFIDENCE

Coordination of information in this way has given us great confidence when it comes to manufacturing information. We won a proposal to extend, alter and refurbish an existing hotel. Our plans included the demolition of an existing timber-framed conservatory along the front of the building and its replacement with a new entrance and glass-fronted link with breakout space and reception accommodation. This new link was to incorporate five different level connections and work with three main level changes across the front of the hotel. The foundations for this building were stepped across the site and the steel frame also needed to accommodate these steps as well as head clearance heights to a tolerance of 20mm when finished.

Our appointment incorporated coordination of all information. Given the crucial nature of getting the levels correct across the building, we worked closely with the structural engineer to provide a coordinated set of (modelled) information to the contractor. He provided us with drawings and we modelled everything. Once the project started on-site, our involvement was limited to coordination; a project manager was overseeing the build.

The coordinated drawings were sent out to tender and a local steel fabricator was appointed. The steel fabricator used CAD-CAM (StruCAD…Steel fabricators were the first people we managed to collaborate with in 3D). They therefore prepared a full fabrication model of the proposed steel frame based on the coordinated information including all connection details, bolts, thermal breaks etc.,which they sent us to sign off prior to manufacture. The fast-track nature of the project, however, meant that this necessarily had to take place prior to the casting of the foundations and setting of the pad stones. We knew this to be a risk.

However, we overlaid their fabrication model on our steel frame model and it fitted perfectly. Minor finish adjustments were made to our architectural model but no major work was required. Coordination was achieved with the Existing Building, the Proposed Extension, the Foundations and the Fabrication model. So we signed off the steel for fabrication subject to the contractor taking responsibility for the rest of the building being constructed as per the information provided.

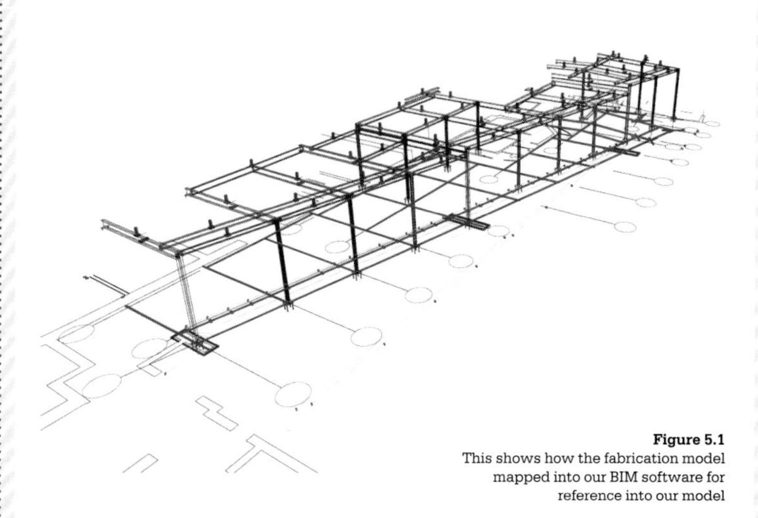

Figure 5.1
This shows how the fabrication model mapped into our BIM software for reference into our model

When we received a call from site stating that the steel frame 'did not fit'. we knew that the frame was correct, the survey was correct and the information provided for the foundations was correct. What had actually happened was that the groundwork contractor had set out the foundation pad heights incorrectly.

A site visit confirmed immediately that this was the case and the problem was remedied via the groundwork contractor, without any blame apportioned to the design architect or supplier. Whilst we always prefer involvement beyond Stage 4, where commissions restrict such involvement, a coordinated BIM approach makes for 'better' information, allows construction errors to be easily identified and reduces risk from the designer's point of view. Moreover, when the inevitable requests for additional schedules, sections and details arise, they can be generated.
For this project we were also interior designers, working closely with the client and fit-out contractor, scheduling everything from case goods to wallpaper. Given we already had the model, we were able to add the necessary detail and use ArchiCAD's inbuilt scheduling feature in conjunction with ArchiSuite's ArchiQuant add-on to do all the necessary take-offs too.

Initially this began as a way to extract their information in the best possible way for us to coordinate. More recently we have found that empowering our collaborators makes for a stronger team and long-term relationship.

For example, whilst working on a project with a local firm of consultants, we instigated a meeting at our offices to run through the critical parts of their information that we had modelled ourselves for coordination purposes.

During this process, they mentioned a number of projects they were involved with where they were contractually obliged to deliver their information through BIM, but were not confident enough in their skills to undertake it.

They invited us to work as sub-consultants on a number of projects, representing them at meetings, checking their models on an ad hoc basis and generally aiding in their transition to a BIM workflow.

'From our point of view we have developed a series of retrofit solutions within our BIM software, refining and adjusting things... to enable even more accurate simulations on future projects.'

THE PROJECT

▶ TECHNOLOGY STRATEGY BOARD RETROFIT FOR THE FUTURE PROJECT

This project came about as part of an initiative to assist registered social landlords to test and monitor differing retrofit solutions to their existing housing stock. The resulting information became part of a larger database that can be used to inform future roll-out initiatives across the country.

We already had an interest in retrofit through our normal practice with more forward-looking clients. We saw this competition as an opportunity to further research our interest in this field, implementing some of the more expensive and intrusive measures which a great deal of our client base didn't have the budget for. We are also particularly interested in retrofit solutions for buildings with a conservation aspect, which can often be cost prohibitive and this competition seemed to have scope to expand more into this area.

The programme comprised a two-stage competition. For the first stage entrants were to propose a range of measures to enable existing house types to achieve the 60% reduction in energy use. We were successful in two first-stage projects.

Using BIM and Eco-Designer, we were able to rapidly and accurately model a variety of retrofit and energy usage scenarios. Our approach to modelling (rather than drawing) – whereby all constructional elements are modelled in the manner in which they are ultimately put together in construction, together with imbuing each element with parametric materiality – lends itself to relatively accurate energy modelling. Whilst this modelling is time intensive at the outset, we were able to amend retrofit measures and reassess the results on the fly.

Figure 5.2
The two completed houses

One of our first-stage submissions was judged to be unique enough for prototyping and TSB selected us along with a broad range of interesting and innovative projects that would give them a wide range of comparable results. Successful projects at Stage 2 were therefore awarded funding towards the measures being installed.

The properties we worked on were a pair of three-storey Victorian terraces previously converted into six one-bedroom flats utilising a shared staircase in one of the houses. They had been unoccupied for some time and were uninhabitable when we came to the project. The proposal was to convert them back into two single-family dwellings, one of which would be detailed to current UK building regulations standard and the other to our enhanced specification incorporating a number of low-energy measures.

During the design process, we used ArchiCAD with an integrated add-on called Eco Designer to analyse the thermal performance of the existing structure (as amended to reflect the updated layout).

Once the baseline had been established the software produced a detailed report for the existing house based on the occupation profile we had created. Taking the existing report as a baseline we then tested a range of options on the property and were able to determine the proposed effects of the measures on the fly using the software. This not only gave us immediate feedback on the options in terms of energy performance, but also allowed us to assess the full impact from a design point of view in terms of, for example, the loss in floor area to internal insulation, as well as the visual impact each of the measures had on the building. Throughout the process we were able to produce a number of different reports detailing the effect each measure had, as well as the combined reduction of proposed energy use for each of the options. The ability to use this work method to back up our whole house approach, together with the detailed information it provided us with, was key to our successful bid for funding on this project.

Our final proposals were to install a range of technologies to the test house to achieve the 60% reduction in energy use.

These included:

- internal insulation to the front elevation
- external insulation to the rear elevation
- super-insulating the roof and ground floor
- solar hot water
- photovoltaic roof slates
- high-performance double glazed sash windows
- airtightness improvements
- MVHR to be fitted within the existing chimney flues.

All of these measures were designed and incorporated within our BIM (as described above) prior to tender stage to allow manufacturers as much information as possible, to keep the prices as accurate as possible. By using BIM to assess the different options, we were able to assess the measures not only in terms of how it would perform, but also in terms of how it would look; whether it would fit; how difficult it would be and how much time it would take to install; and of course, how much it would cost. During the design process, we also looked at things like heat pumps and alternative methods of achieving the required airtightness targets. Given this information, we were able to come to a valued judgement far more efficiently than would have been the case which, given this was a competition and the work was – initially at least – at risk, was probably the difference between us entering the competition or not.

The most critical area for coordination of the above was the MVHR system within the existing chimneys. We had an incredibly accurate model of the existing building and were sure the ductwork would fit in the space allocated. We issued the documentation for tender with an indicative layout, a full set of plans/sections/elevations and an IFC file. So when one of the suppliers sent us a competitive quotation with 3D drawings of the system overlaid on our drawings, we were heartened.

We obtained the model from them (dwg format) and referenced the geometry into our BIM. It didn't fit and the component list was short by about 40% of the materials required to complete the installation.

The point to take from this is that all of this was picked up prior to the order being placed and prior to work commencing on-site. Resolution of this issue was simple enough. We amended their 3D model, sent it back and had them prepare an amended quotation based on the new information. When the equipment arrived on-site it fitted and there was no waste and no delay whilst any additional parts were ordered. In short we were saved the three week lead-time on-site, a potential cost increase and a couple of night's sleep. We were able to convey, in great detail, the mechanics of the project to both client and contractor, optimising costs at tender stage, streamlining scheduling and minimising site waste.

On this project the BIM workflow allowed us to win funding for the project but it also helped us at every stage with our design, energy assessment, specification, communication, promotion, coordination, listing and documentation and with minimising waste on-site.

'We use BIM across all of our projects – from house extensions through to 80-bedroom hotels on both new and existing buildings'

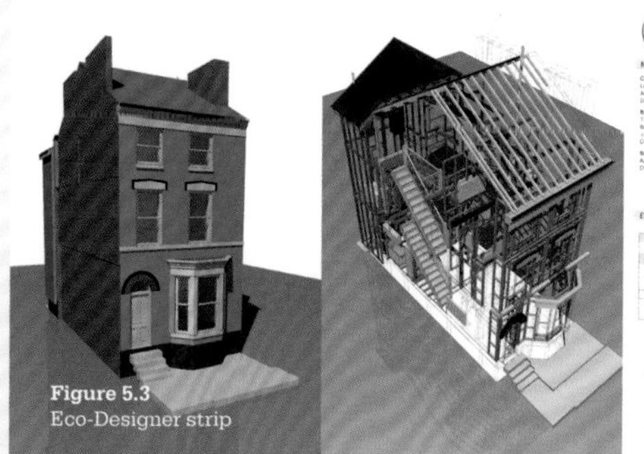

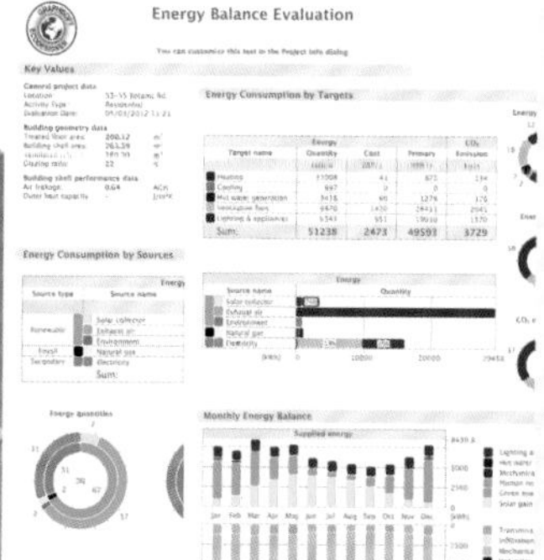

Figure 5.3
Eco-Designer strip

CONCLUSIONS AND FUTURE PLANS

The lessons learnt from this project are wider than the narrow sphere of our practice and BIM. As discussed above, the project is being monitored to test the effectiveness of the technology installed under the scheme. From our own point of view we have developed a series of retrofit solutions within our BIM software, refining and adjusting things like cost and installation time and efficacy further to enable even more accurate simulations on future projects.

All of this information is stored on our various templates within the office and can be harnessed where appropriate on future projects just as it was on this one. Like all models the information you put in has to be as accurate as possible to get a realistic comparable result. However, the main benefit to our practice through this process has been in terms of proving the validity of our approach and the benefits of using BIM at every stage of the process to arrive at the correct solution as efficiently as possible.

Using BIM within the practice has enabled us to develop a very reliable, robust and consistent workflow. It's not perfect yet (and will even evolve, we suspect), but we feel our BIM approach gives us the freedom and peace of mind to operate with confidence.

Where possible we will always try to work with other consultants collaboratively and have in the past done so successfully using both native file formats and IFC. However, many other consultants for smaller projects have a lot of catching up to do. There are many technical and economic benefits for smaller consultants to get on board and we encourage others wherever possible.

We're in the position now where we've started to get a reputation as experts in this field and have even been asked to take on work as subcontractors to larger organisations, but on the projects we work on for ourselves, our BIM is largely a lonely activity!

At the beginning of the chapter we alluded to the architect's traditional role being that of coordination. However over the years this has been eroded by other professions as the caricature of the ego-driven architect who makes things expensive and delivers projects late has been allowed to take over. Part of this has been the inability to fully embrace the more responsible and profitable role of coordinator over the more recognisably cool role of designer.

BIM should be seen as an opportunity to bring some of these responsibilities back into our profession, bringing us neatly to a new role that is emerging: the BIM coordinator responsible for putting everything together and ensuring nothing conflicts. Wouldn't it be great if architects could take on this role and make sure not only that it works, but also that it's the most elegant and efficient solution possible?

'A coordinated model allows us to spend more time on design, allowing us to assess, document and present a wider range of options in a shorter period of time'

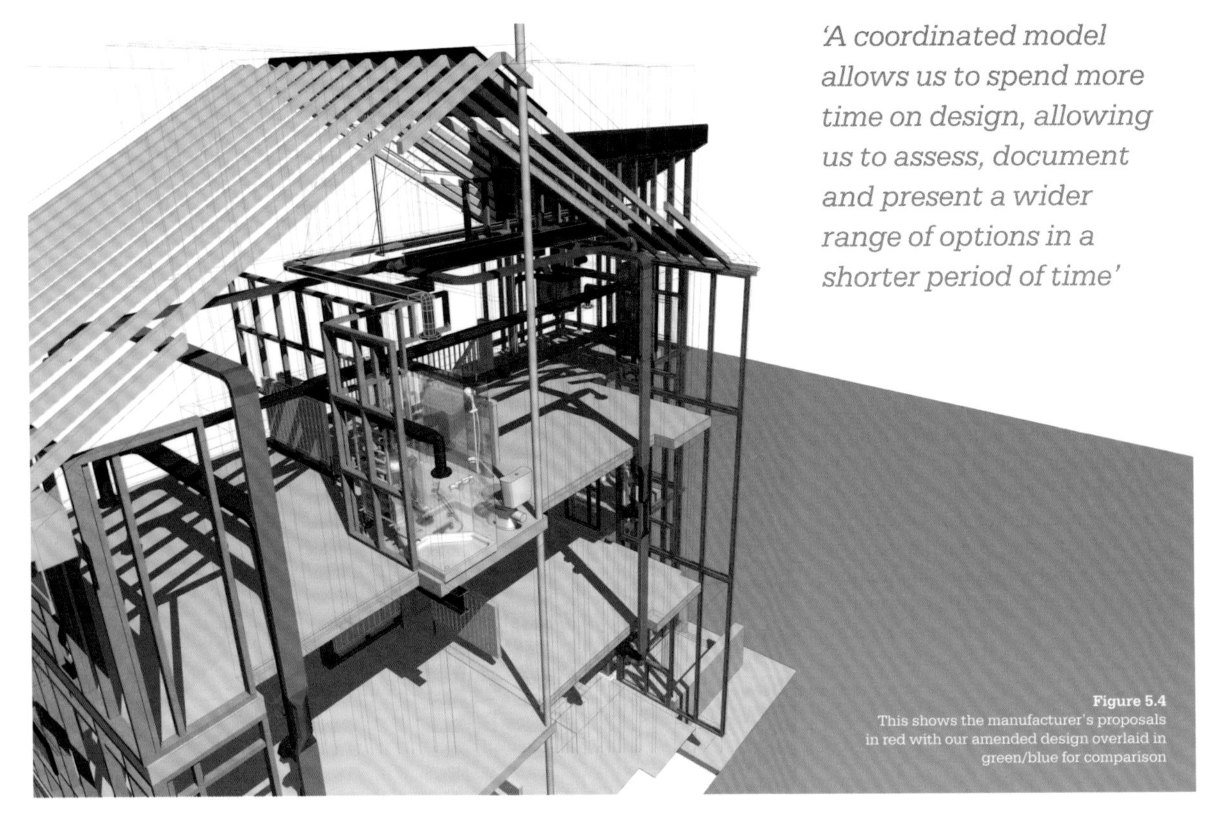

Figure 5.4
This shows the manufacturer's proposals in red with our amended design overlaid in green/blue for comparison

Our practice is actively seeking opportunities to work on larger collaborative projects where BIM experience is required. We don't see collaborative BIM work as drawing other people's projects; rather the early involvement that BIM requires encourages collaboration. This is one area where BIM can work for smaller practices.

Implementation for a large organisation is a massive outlay in terms of training, licensing and hardware, whereas a sole practitioner with enough experience and around £10k outlay can set up as a BIM coordinator in addition to their current services. Architects are already good at this, they just need to make the jump. Perhaps then we can take back some of the more profitable parts of our role which seem to be carried out by other consultants, but which architects should be uniquely placed to perform.

Figure 5.5
This is one of the first projects where we used BIM to coordinate information. It was completed in 2005. We still use it to illustrate the power of the approach

LESSONS LEARNT

We would offer the following three tips for other practices beginning to use BIM:

- Without a decent understanding of construction BIM is next to useless. Knowledge of design and building construction is much more useful to us when recruiting than being able to use any particular software package – the former takes five years at university and two years of practical training as a minimum, the basics of the latter can be taught in a matter of weeks. For this reason the training of staff should begin with project architects rather than Stage 1 Architectural Assistants.
- Select your software carefully and try to work towards Open BIM where possible. Accept that no one software provider can fulfil the needs of the entire construction industry and that other stakeholders – be they clients, contractors, engineers, surveyors, whatever – may need to use something different. Specifically, whichever software you use to implement BIM, adopt the AEC (UK) BIM Protocols as soon as possible. That way when you do start to collaborate, it should be a much smoother process. People who have been doing it for years have written them and the documents are free. Use them!
- Don't forget to account for some of the more hidden costs of adoption. Time spent learning BIM, where it is not related to a specific project, is research and development and should be accounted for as such by the company. Additionally, in the spirit of our case study, look out for funding streams to support the transition period. Whilst there may not always be specific money for training, there is often match funding for some of the related activities, which can be accessed if you are prepared to look and think constructively.

MOVING TOWARDS BIM IN THE CLOUD

Johnathan Munkley

Practice:
Niven Architects

Location:
Darlington

Employees:
4 architects, 10 technicians, 3 admin staff, including a practice manager and PR/marketing officer

Founded:
1984

Director:
Simon Crowe

Technology:
Revit; Navisworks; Tekla; Autodesk Showcase; AutoCAD; ArchiCAD; Vue Xstream

06

This case study focuses on using the increasing number of BIM tools that provide access to tools in the cloud. Cloud computing, which is already prevalent in other fields –notably visualisation where scenes are uploaded and processed cheaply using massive computing power – is starting to make an impact on the world of BIM with solutions for online collaboration, analysis and live modelling. This case study centres around a Technology Strategy Board funded programme to review the potential and develop protocols for cloud- based working within the design team.

THE PRACTICE

I have been working for Simon at Niven Architects for six years, having started after completing my Architectural Part 1 on a 1-year placement. I had some knowledge of BIM from my first few weeks at university in 2005, where we began using ArchiCAD after a very inspiring lecture. As a firm we operate in many sectors but particularly education, retail, commercial, industrial and leisure.

It is vital for any firm implementing BIM to have a sector-wide understanding of various vendors' software tools and, more importantly, the workflows that are enabled by the different technologies. The first task was to gain a thorough understanding of all the available technologies, as well as understanding and developing our business strategy before committing to a large capital outlay. Our decision was made based on what we could afford and what was technically best suited to us. One of the difficulties associated with this decision is that in any BIM project there is an array of tools utilised by various design team members, external consultants and contractors.

In the future this will expand, inevitably, as an increasing number of stakeholders, such as those in the second tier of the supply chain – manufacturers and part suppliers – become involved with the BIM process and more BIM technologies emerge. Practices need to be able to solve issues relating to multi-vendor projects and keep up to date with emerging technologies to ensure the success of any given project.

We actually started using 3D in AutoCAD and moved to SketchUp in 2006, but we soon realised that BIM could bring much to our practice and also improve our bottom line performance and commercial outputs and so we decided to invest in basic BIM in 2006.

We bought the first Revit licences and started the implementation and training processes. However, due to unstable software functionalities at that time, a skills gap, the recession of 2008 and the lack of appreciation of the full implications of BIM on the practice, the implementation process took longer than expected.

In 2009 I gained experience of working on collaborative live BIM projects in Australia where I was employed to coordinate information from a design team after they fell out. It was a very strange situation but the benefit of fully coordinating design teams' information was obvious from the outset. It became clear to me how much of an impact BIM could have in our practice and supply chain and the potential benefits we could obtain from thorough implementation. This would distinguish us from numerous small practices where BIM is utilised only for front end design.

In September 2010 we teamed up with a research team at Teesside University's Construction Innovation Department, composed of Professor Nashwan Dawood and Dr Mohamad Kassem and under a Knowledge Transfer Partnership scheme funded by the Technology Strategy Board began working to transform our practice. Through this partnership, which is still ongoing, we have established protocols for an integrated BIM project delivery.

This puts us in a good position for the Government Maturity Targets (file-based collaborative BIM sharing) by 2016. After a year, despite several of our projects being started in a non- BIM environment before this full-scale implementation project, more than 50% of all our projects are now using integrated BIM and more than 75% of projects will be using it by the end of 2014. We felt that the integrated BIM approach would bring the most return on investment for us and our project stakeholders.

During this transformation to BIM, we have appraised the commercially available BIM technologies, reviewed the BIM protocols and workflows developed for the UK (i.e. AEC (UK) BIM protocols, CIC BIM Protocols), developed our own protocols for integrated BIM project delivery and implemented CPD and training for our staff.

This consisted of all members of staff in 2008 being given a three-day Introduction to Revit course, to give them the basics of how to author a building. However the course cost £3,000 and it was therefore obvious that this wasn't something we could run every time a new member of staff joined us or when we wanted to develop skillsets.

We initially encouraged staff to undertake self-learning in their spare time, modelling projects they had done in the past to help develop skills. We ran CPDs in house that BIM champions were tasked to develop. Some of these consisted of a basic software introduction (for example Autodesk Showcase or Navisworks) whereas some focused on BIM knowledge transfer so staff could gain a better understanding of BIM. This allowed for skills to be transferred down our BIM hierarchy.

Although our staff now have expertise in a range of tools our main authoring software is now Revit. Our internal review of available BIM technologies revealed that Revit can meet most of our specifications. Due to our existing subscription to Autodesk, this meant it would have been a financial burden to go down any other route.

We have developed our bespoke workflows to share central building information models, using various collaborative online cloud-based systems such as Dropbox, four projects, Asite and Riverbed. All design team members including external consultants connect live to a project database, giving them the ability to access live and up-to-date information. Historically we have used Tekla for clash detection and model coordination.

We believe this is a good option for firms starting out due to its simplicity and solid outputs. We have now started to use Navisworks for model coordination and checking and various other tools as and when they are required in projects. Each project may require different BIM deliverables and processes and therefore necessitates a different set of tools to achieve this.

Being an SME, we faced two challenges during our journey to an advanced level of BIM implementation. The first challenge was our limited capacity and power to influence the consultants and other stakeholders of our supply chain. The second challenge was related to the 'limited' resources and knowledge required to systematically implement BIM processes and migrate from traditional CAD paper and file-based sharing to a working, live modelling environment. For the former challenge, we advise an open, honest, incremental and collaborative approach with all of your company stakeholders to help them appreciate the benefits and start to move to the next step on their BIM journey. For the latter challenge, seeking assistance from experts such as academics that could be accessed through particular government-supported grants worked well.

'The first challenge was our limited capacity and power to influence the consultants and other stakeholders of our supply chain. The second challenge was related to the limited resources and knowledge required to systematically implement BIM processes and migrate from traditional CAD paper and file-based sharing, to a working, live modelling system.'

THE PROJECT

The project we've chosen is the £1.4 million residential development for the Camphill Trust at the Larchfield Community, Hemlington, in the north east of England. The residence contains nine self-contained apartments with communal living spaces, internal landscaped courtyard areas and individual garden areas.

The first challenge on this project was the implementation of live data sharing technology and the development of protocols. Most firms are not in such a strong position as us regarding this, as we have benefitted from the help of the University of Teesside, who have been fundamental to our success in overcoming this first challenge.

The second challenge was to convince the design team that it was the right approach to deliver the project and potentially adopt it on future projects.

Figure 6.1
Initial concept image for Camphill Trust development

Figure 6.2
Initial concept image for Camphill Trust development

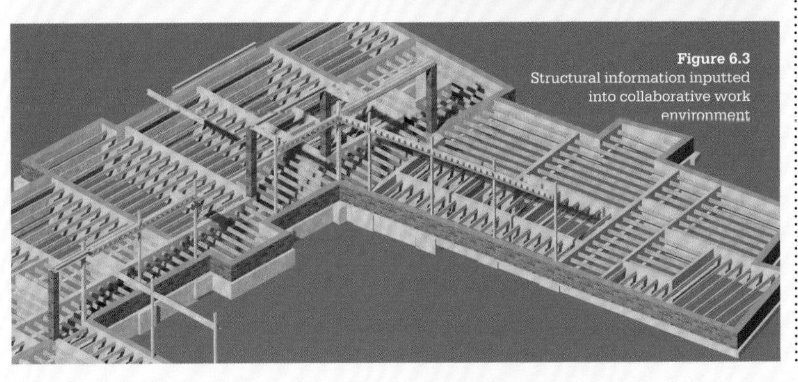

Figure 6.3
Structural information inputted into collaborative work environment

Being an SME working with groups of consultants, doing the hard sell on BIM has become a part of our daily routine. We have spent considerable time trying to convince consultants to invest in BIM technologies and offered our support to help them understand the new processes. At the end, we would not be able to keep our working relationships with consultants who are not 'BIM enabled' as our processes would not be compatible. One final challenge is related to the investment in IT hardware of our stakeholders. We have invested heavily in IT infrastructure to implement BIM processes without being restricted by computer power.

To cope with the limitations posed by some underpowered computers utilised by some consultants, we had to develop protocols to give staff instruction on how to issue work to stakeholders with 'slow' computers. These 'additional' workflows are not needed and could be saved if the IT infrastructure of stakeholders is upgraded and is compatible.

A multidisciplinary integrated model was developed using the BIM live modelling system. This project was utilised as a pilot project to test the processes and protocols we had developed. The model was established in the cloud and consultants were trained on the processes and protocols to tap into the live data. The ultimate goal was to get all our design team, including the client, interacting with the singular integrated model. Once the live modelling system was up and running, subsequent processes were more traditionally architectural, such as producing planning and tender documentation. It was impressive to spot and flag design mistakes as consultants' information was appearing via the cloud in the live model!

LIVE MODELLING AND A DATA SHARING ENVIRONMENT

Central data sharing was vital to the success of BIM implementation within our practice. Some staff had prior experience in Australian design teams using the ArchiCAD BIM server, so had seen the benefits of this implementation level.

Despite the fact that they were dispersed geographically, using only ArchiCAD and Skype, they felt as if they were working in the same office. We believed that working on a central shared model would suit the nature of our business, as our designs vary across projects and are based all around the country. We concluded that a live central data environment was the ultimate goal for us to get the most out of the BIM approach.

The suggestion to use a live modelling environment garnered mixed opinions. Some could not grasp it. Others felt it was a risky process, concerned that they would not be able to check work before it was 'issued' if the live environment made designs immediately available. The idea of instantaneous live modelling bypassed all of the traditional quality assurance systems, which was a problem for firms that were ISO 9001 QA certified.

However once I explained that drawings would still be formally issued at specific stages to become legal tender, the firms involved were open to the idea. It is understandable that they were nervous of using such a system, as it represented a fundamental change to traditional workflow.

Live modelling is just a small step in a much bigger picture of BIM implementation, so it was vital to us.

After a few design team meetings, using examples from pilot testing, the advantages of central live modelling were understood by most of our supply chains.

As we watched the pipework appear live in the ceiling, clashing with other building elements, we realised that the M&E consultants had misinterpreted a bulkhead design detail which we instantly flagged up and they amended the model. Without the live modelling this mistake may not have been spotted until site or would have resulted in aborted works and redesign.

Even though design issues were flagged very quickly it still took time for consultants to make any changes, as they had generally moved onto different sections of the model. However, once they received notification, mistakes were rectified and this still reduced the number of aborted works.

Because our design teams work from various offices, issuing information and collating varying file formats had always been an issue that caused aborted works and redesign effort. So creating a collaborative working environment was vital to what we wanted from BIM. The greatest impact of this approach is its ability to compress design lead-times. Working in this environment according to predefined protocols, the traditional 'sequential' design sub-tasks tend to overlap and the whole project benefits from a reduced duration.

This was the most difficult aspect to communicate to consultants and other stakeholders, as it represents a paradigm shift from their traditional way of operating in company silos.

Some of our consulting design firms have been using BIM-based 3D authoring platforms but they have not looked beyond their office boundaries for its use and benefits within the collaborative working environment. These reduced our design lead-times and aborted works and the initial results on RFIs are also good. We are looking at the cost savings for the overall project based on BIM-based protocols, however this will take time and is difficult to project. Working with Teesside University Construction Innovation Department allowed us to explore this externally to project work.

Working in a BIM live central model is still suffering from challenges such as trust in other people, data and information and intellectual property protection, especially in the absence of clear legal frameworks that address these aspects.

People were concerned that the data they were inputting could be manipulated by other consultants and therefore placed risk on them, however we remedied this by locking information into sets so each consultant could only manipulate their information and we robustly stated liability for information within the BIM execution plan and the single fact that only traditional formally issued drawings at specific stages would be legal tender.

Figure 6.4
Problem with the live nature of the collaborative environment

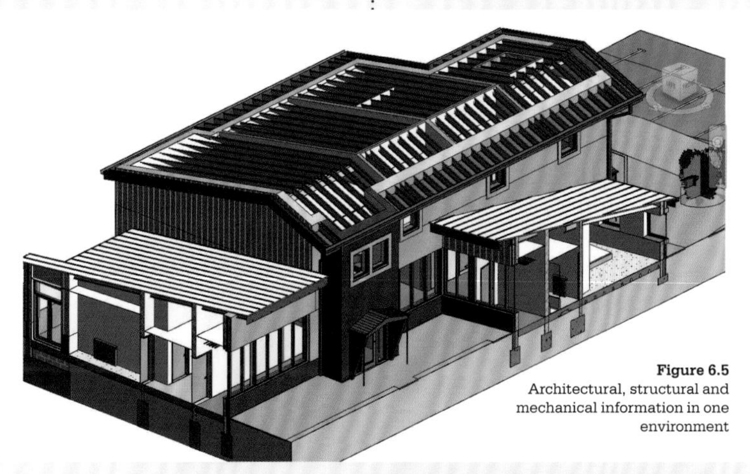
Figure 6.5
Architectural, structural and mechanical information in one environment

Figure 6.6
Architectural information that has been inputted into the BIM

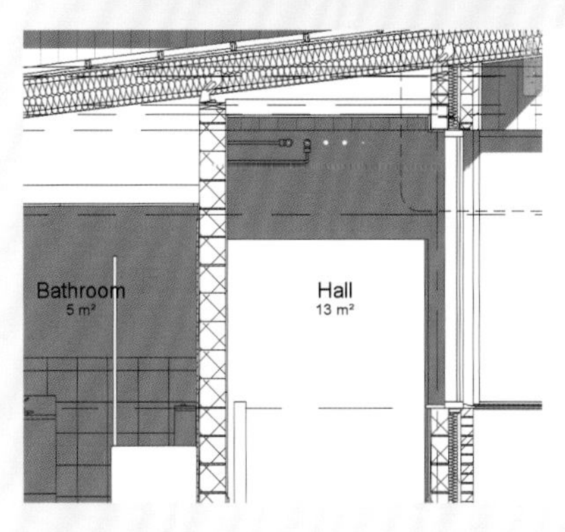

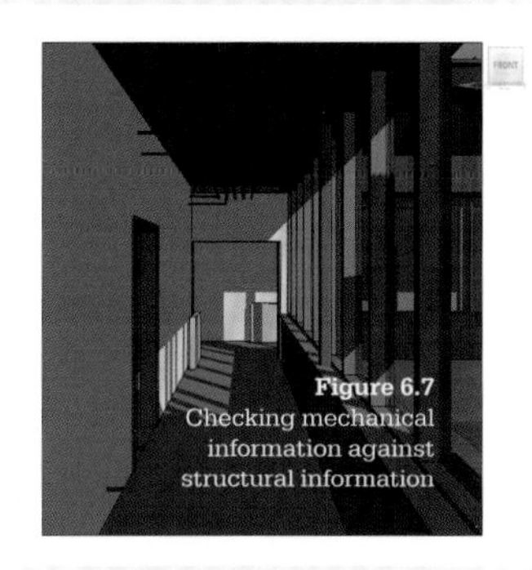
Figure 6.7
Checking mechanical information against structural information

It was the first project we used the BIM live modelling system on – a system which we developed that allows all of the design team stakeholders to input and extract data from one model direct from a cloud-based solution, with everyone producing architectural information in one place at the same time. It was an ideal case study as we were the lead consultants and the design team and consultants were open to the methods we were proposing. The size and scale of the project proposed do not entail a high level of risk for testing new processes. The design brief was not too complex and therefore we could focus on testing the processes, which is key at this early stage. Also despite the fact that BIM was not required by the client on this project, we felt it was a good project to introduce new processes to our teams and consultants without causing risks to any of the parties involved.

The two main outcomes consisted mainly of a successful test of our BIM live modelling system and our consultants buying into the new processes of live modelling. However, there was resistance from some parties as they were worried about inputting into a central mode and data set on a real time basis when information is still in progress and therefore has not yet been checked. While recognising these challenges, our practice intends to progress working with central live models as it fits our business model. On this particular project, its adoption was a success and we are currently using it on much more significant developments. We believe that BIM live modelling will become a vital aspect of any BIM project in future as emerging BIM technology and industry testing would allow an understanding of both benefits and processes.

'Creating a collaborative working environment was vital to what we wanted from BIM. The greatest impact of this approach is the ability to compress design lead-times.'

CONCLUSIONS AND FUTURE PLANS

Our ultimate objective is to deliver buildings to clients on time, on budget and according to the agreed statement of work. This is the main reason for our firm's existence. BIM does change the methods for delivering projects and in particular the way and timing of producing data. We now have a longer and a more intensive front-end process, which allows us to produce better design before moving to tender and site. For example, our conceptual design stages now receive input from all project stakeholders from day one.

This allows us to value-engineer at this early design stage and save on aborted work further into the project, therefore creating efficiencies. Despite the benefits of this change on downstream processes such as detailed design and site, many of our existing clients are still trying to request work to be completed in short timeframes. This is obviously detrimental to our BIM workflows.

However, we expect that this will change in future and clients themselves will be asking for BIM projects as they gain awareness of its benefits to their projects. Moreover, there could in future be some shift of fee from construction stage to design stages to compensate the increased effort at the earlier design stages. In some countries like Singapore they have already pursued that direction. We hope this will happen here in the UK too, in the near future.

The biggest impact of BIM has been on our business model. We started to position the company during the very early days of the recession to be a technology and design-led architectural practice. This has allowed us to target projects that were historically out of our reach. We have been able to move from small refurbishment projects towards targeting larger and more profitable projects and have plans to start targeting BIM-driven projects in emerging economies.

We have already gained international recognition, winning an award at the Open BIM Build Qatar Live and Build Sydney Live competitions, so we are hoping to transform this success into new international project work. We also work on government-funded frameworks.

With BIM, once the initial models have been produced, it is much easier for smaller design teams to make amendments to massive projects, as the parametric and integrated nature of BIM technology facilitates the execution of many tasks. We are being approached by large contractors looking for BIM-compliant architectural practices to the extent that we had to reject work from contractors we had hoped to work for.

We have strived to select technologies that fit our 'mapped' business process requirements with their standard off-the-shelf features. We have invited software vendors to give live demonstrations in workshops to our design team. However, one size does not fit all and some customisation will be required. In many cases, the customisation does not necessarily entail modifying the technology itself, instead it could be achieved through developing tailored workflows or protocols.

This was the case with our live central modelling. We have utilised software which included off-the-shelf authoring technologies and a web-based sharing technology and designed our own IT hardware infrastructure, and more importantly enabled such software and hardware infrastructures to work through the deployment of tailored workflows and protocols (naming convention, saving location, etc).. In other cases and for elaborated design decisions, a bespoke development might be required.

Our day-to-day interactions with consultants have obviously changed with these new BIM workflows. A simple example of change, which resulted from the cloud-based technologies for central data storage, is the saving of time spent on sending emails and data storage devices. More importantly, we interact virtually and more regularly with our consultants to coordinate BIM information, which cut down meeting times. The major benefit of this is having design team meetings and inputs as and when we require them rather than on set dates.

'The biggest impact of BIM has been on our business model. We started to position the company during the very early days of the recession to be a technology and design-led architectural practice.'

LESSONS LEARNT

- Business model first and technology second: one of the mistakes SME firms can make when they consider adopting BIM is that they buy an authoring BIM technology and then think about what they want to achieve with it. The business model and requirements must come first and drive not only the selection of the technology but also the whole process of BIM implementation.
- Senior management support is the key to successful BIM adoption: the lack of senior management support for BIM adoption is the perfect recipe for failure. Senior management must provide the resources and tools, give staff time to go through the learning curve on their first projects and tolerate mistakes during this transition period as part of the risks involved.
- Appointment of an experienced BIM champion: during the early stages of BIM adoption an office BIM champion is vital. He can be the 'go-to man' for all technical queries of the practice BIM production team, saving much time and improving work efficiency. He will also be able to develop BIM protocols and workflow and train staff, allowing your firm to get off to the right start. BIM adoption can be costly and the last thing a firm needs is to set off down the wrong route.

USING EXISTING BUILDING BIM

Robert Klaschka

Practice:
Studio Klaschka

Location:
Southwark, London

Employees:
1 designer, 1 architect and 3 architectural technologists

Founded:
2001

Director:
Robert Klaschka

Technology:
Bentley AECOsim Building Designer; Bentley Pointools

07

Working in existing buildings has for a long time been outside the intended scope of BIM. In this case study the revolution of modern lower cost methods of surveying and modelling is discussed in detail. Centred around two health projects in East London it discusses the process of laser scanning and dealing with the massive amounts of data that point cloud surveys represent and modelling buildings from the point cloud. Combining the point cloud and the model the study shows the benefits and economies that come from this workflow.

'When we set up our own practice we had no legacy systems or staff training issues and so decided that we would pick the best design software we could find.'

THE PRACTICE

When Studio Klaschka was set up in 2001 my then business partner and I had just left David Morley Architects (DMA) where I had, amongst other things, advised on the direction the practice should take with technology. When we set up our own practice we had no legacy systems or staff training issues and so decided that we would pick the best design software we could find.

We had experience of Microstation at DMA and had started to look at Triforma – which was and still is the foundation of the Bentley BIM software products. Since then we have moved onto Bentley Architecture and now use Bentley AECOSIM.

This represents the evolution of the Bentley products over the last 12 years. We began looking at point cloud survey information in 2007 because it provides much more complete information than a traditional 2D CAD survey. At that point it was not practical for us to start using the technology for several reasons. There was no simple way to integrate it with the BIM software that existed, the software that was available was prohibitively expensive for a small practice and commissioning scans also had a high cost because it was not yet widely used. The market was limited to higher value work and was only considered for specialist architectural uses.

This all started to change in 2010 when Bentley initially licensed and subsequently bought Pointools, which is considered one of the best and fastest point cloud engines on the market. They incorporated the engine into their BIM authoring software, which we use and so we effectively got point cloud functionality inside our software for no more than our annual subscription costs. Around the same time Faro introduced the Focus3D. This is a smaller, cheaper scanner than most on the market and works well, especially for scanning the interiors of buildings which often require between one and three scans to capture a whole room. The price of the Focus3D has enabled lots of smaller surveying companies to start laser scanning and so the market has developed.

Since the recession hit at the end of 2008 every job we've had has involved refurbishment or extension of an existing building and as a result we have had 15 projects scanned between 2011 and 2013. These scans have formed the basis for our models. They are referenced directly into our software and then traced.

Recently we have been working on small health refurbishments for East London and City Alliance (ELCA), which oversees the estates for City and Hackney, Tower Hamlets and Newham PCTs. Over the last three years of funding rounds our work has included aesthetic improvements, replanning and refitting health centres and day patient wards.

The projects in our case study are our most recent and both take advantage of point cloud survey data that we commissioned and that was used to produce an existing building model as the starting point for our design and production information.

THE PROJECTS

Initially we started working on these projects using supplied traditional survey information, but it quickly became clear that the information was not really detailed or accurate enough. The main consequence of this was that we had to return to site frequently to gather new information that was missing from the survey.

Since then we have had five health projects point cloud surveyed using two different surveyors. In some cases the project architect has never had to attend site because the dimensional information from the point cloud coupled with the picture panoramas that are also captured and accessed through a web portal provide very complete information.

These schemes include those below and a roof that was scanned as part of a record-keeping analysis exercise on a large Grade 2 listed health centre, a compliance checking exercise for a dental floor in a new health centre and a levels and gradient review of an access route across a hospital site to an nearby bus stop.

▶ SORSBY MEDICAL PRACTICE

This small building in the London Borough of Hackney contains a little GP practice on the ground floor and administrative offices for the Homerton Hospital Services, including health visitors and district nurses. There is also a single, small, undersized treatment room on this floor. The building appeared to have been heavily modified, but had limited scrutiny for both spatial standards or infection control. The exterior of the building is in a poor state and has no architectural coherence; it also suffers from overheating in warm weather.

Our initial studies confirmed that a small extension would be required to make the GP rooms of an adequate size to meet the HBN guidance and we also took advantage of the extra space to bring all the treatment spaces onto the ground floor to facilitate ease of access and remove the need for patients to go to the first floor.

Figure 7.1
Sorsby Medical Practice model including plans to extend the building and add solar shading

▶ WEST HAM LANE HEALTH CENTRE

This existing health centre building in the London Borough of Newham has been the subject of three phases of work. Initially we brightened up the envelope. Then we did a series of technical fixes including replacement of the existing boiler and resolution of some roof access issues.

The final and most significant phase focused on the ground floor where there were significant problems including poorly utilised space, duplicated waiting areas and lack of visual scrutiny of public areas.

There were also problems with the fire strategy, which related to the time that the building had been constructed and so our design introduced protected lobbies to the main staircase. The building had to remain in full operation while the work took place and this introduced a requirement for complex phasing.

Figure 7.2 West Ham Lane Health Centre after phase one to refurbish the cladding

'I've found that we, as architects, are ideally placed to economically model the existing BIM.'

▶ WHO SHOULD MODEL THE EXISTING BIM?

I've found that we, as architects, are ideally placed to economically model the existing BIM. Having often looked at going down the route of commissioning the existing building model from a surveyor instead of producing it ourselves, time has told that this isn't the best route to take for a number of reasons. Scanning, registering and then building a detailed existing building model is a lengthy process and can take 3 – 4 weeks even on a relatively simple project. Given the rapid nature of design programmes this starts to cause real problems because it means that either the front end of the project is produced with inaccurate information or has to wait until the survey arrives. It is also difficult before a scheme design has progressed to know which parts of the building need to be modelled in detail. For example on West Ham Lane, when we received the point cloud survey we modelled in-house in a much more parallel way and so cut down on the time required. We began by modelling massing and floor layouts quickly and the design team was then able to use this to progress option studies and initial discussions with the client while the detail was gradually added to the model. At the point that the layout was frozen the more detailed model had been completed. There were even situations in which the point cloud becomes the main source of reference, with only small local areas of modelling or none at all being necessary. An example of this was when we created a new accessible route at St Leonard's Hospital. We modelled a new access ramp with all other checks being made against the point cloud alone.

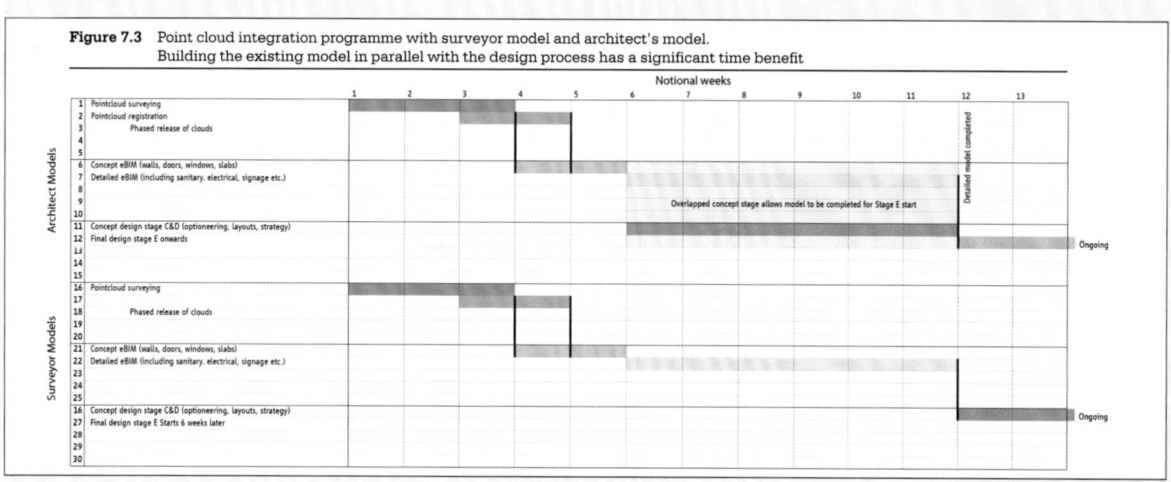

Figure 7.3 Point cloud integration programme with surveyor model and architect's model. Building the existing model in parallel with the design process has a significant time benefit

▶ THE POINT CLOUD WORKFLOW ON THESE PROJECTS

The point cloud surveying of the two health buildings had additional benefits above the normal capture of a very complete set of geometric and photographic data. In this instance, because the buildings are both in occupation throughout the design process, access is more restricted. This reinforces the benefit that with a point cloud survey there is no requirement to revisit a site.

Commissioning a point cloud is similar to a traditional 2D survey. It is only the deliverable that is different. The scan itself will render with intensity. This produces an image in which pale surfaces appear as lighter greys and dark unreflective surfaces appear as darker greys.

However in addition to this colour photographs can be captured too. The colour from these photographs can then be mapped to the points by the surveyor using software. This enables a full colour rendition of the scan to be displayed.

This is a very good medium for exploring the scheme and can be used to create animations and stills showing the full detail of the building.

It is important to remember that laser scanning is a line-of-sight technology. If you can't see something it won't be captured. That could mean roof voids, risers or external roofs and surfaces that cannot be seen from the ground or perhaps cannot be captured during the initial survey. For example on West Ham Lane the centre is in permanent use and the demountable ceiling could not be taken down to reveal the services above.

During the design the services engineer made assumptions, which we planned to follow up with a further localised survey at the beginning of construction to allow the design to be completed. It is important that you consider what you need to know and discuss with the surveyor how roofs can be accessed or if there are places that overlook roofs. There are some ways to get around this with lower buildings.

For example, surveying with the scanner mounted on a tall pole. These vary between surveyors. The largest I have come across is 12m tall and may affect the cost of the surveying considerably if one can survey from the ground where another needs to get on the roof itself.

Line-of-sight also means that if a room is full of equipment then a lot of the room may be obscured. It is preferable to scan buildings when they are empty, but this was not possible for most of the health schemes we've had scanned, because they are usually in use.

The other useful output that most surveyors will provide is a web portal that displays the photographic information that has been captured. This allows any member of the team, even those without point cloud software or hardware, to run it and see the surveyed data. These appear as panoramic images that can be rotated and zoomed at each scanning location. These are also useful for identifying smaller features such as switches and sockets that don't appear as clearly in the point cloud data itself.

Figure 7.4
The scanner first captures intensity, which is similar to the reflectivity of the surfaces. It can then additionally take a dome of photographic information which can be mapped to the points to provide full colour.

Figure 7.5
The photographs captured during the scan can be hosted and viewed using a web browser

'This is a very good medium for exploring the scheme and can be used to create animations and stills showing the full detail of the building.'

▶ ANALYSING THE EXISTING BUILDINGS FROM POINT CLOUD DATA

One of the most useful aspects of a point cloud survey is that it can be located both in its plan location, orientation to north and its vertical height. This means that by inputting correct longitude and latitude it is easy to review daylight and shadow casting in a given situation.

Furthermore the height location means that drawing conclusions about floor levels for flood risk assessment is simply a case of taking measurements of height above the file origin.

In order to achieve this the surveyor ties the survey positions to a series of control points called a total station that in turn are located in the real world coordinate system of your choosing. In the UK this is usually OS Datum. Doing this makes it very easy to tie the position of the other geo-coordinated data such as OS superplan or proprietary 3D city model data for larger area context. It also makes positioning the building as required in a BS1192 project space statement very straightforward.

We generally ask for our scans to be delivered offset on the XY plane to the nearest 10m from the file origin because this makes sharing information with other consultants on different packages easy.

One very quickly realises that existing buildings aren't as square or as flat as you'd expect them to be from the very orthogonal designs that architects often draw. Construction tolerances, subsistence of floors and inaccurate setting out all lead the building to be a much less predictable environment to model than a new building.

Modelling packages, however, expect walls to be straight and floors to be flat. We get around this by looking at the point cloud and deciding on averages. It is important that this decision is supported by all information about the building. A good example of this is to supplement a flat modelled floor with a colour by elevation rendition of the floor including some level points to give reference.

On West Ham Lane when we converted the first floor to an open plan area this revealed that the floor had been built up with a shallow ramp and had a level difference of over 50mm on either side of some walls that we were removing.

Referring back to the point cloud like this to examine areas in detail where there is tight spatial coordination is a common exercise. It is another example of where a designer's judgement can affect the level of detail of the model.

A room that is being painted needs far less accuracy and detail modelled than a space between two walls that a piece of joinery must fit. We keep notes to be read with the model and visual aids that explain the assumptions we have made.

▶ USING THE POINT CLOUDS

On both projects once the scan was delivered it was simply referenced into our BIM package. At this point seeing the walls within the building is not straightforward because the entire building is visible, so we clip the cloud until the section of wall and the floor it stands upon is visible. In order to begin modelling we also need to define a floor plan on which to draw the base of the walls.

At this stage we clip the cloud further and begin modelling the layout of the building by tracing to the point cloud and then widening the walls to their true widths.

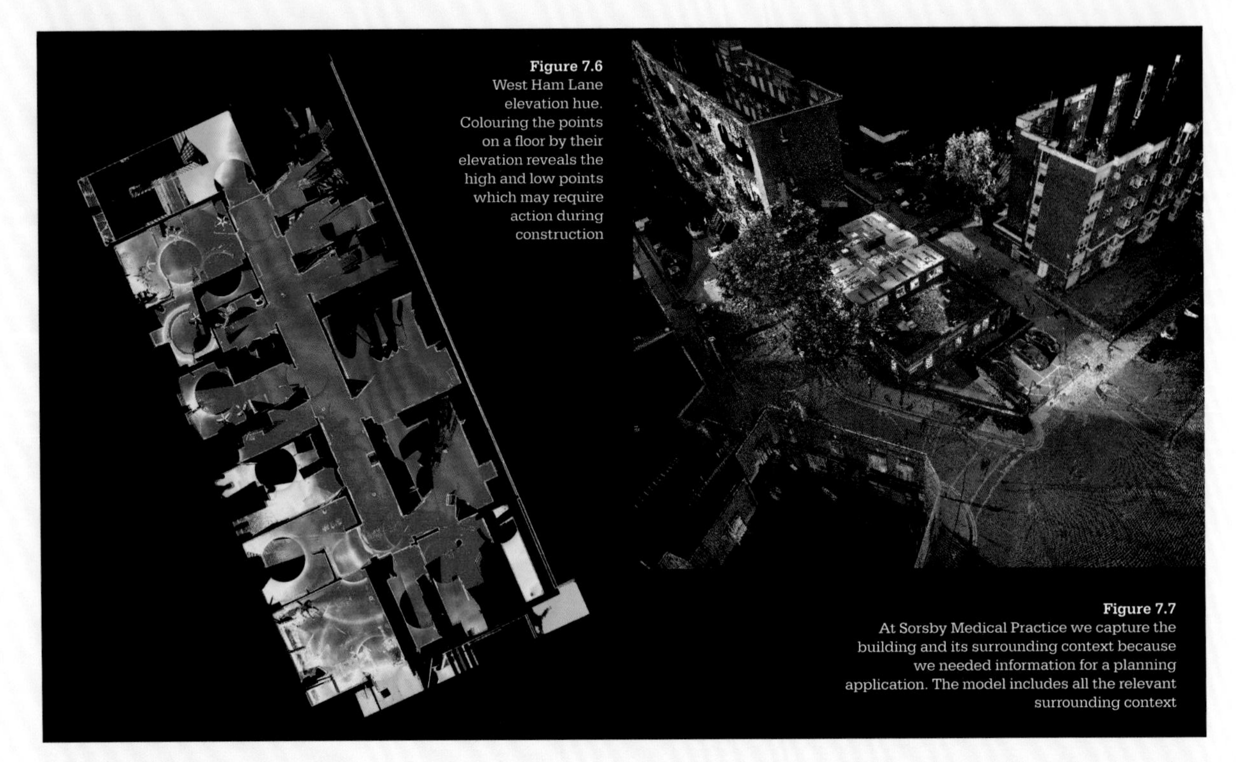

Figure 7.6
West Ham Lane elevation hue. Colouring the points on a floor by their elevation reveals the high and low points which may require action during construction

Figure 7.7
At Sorsby Medical Practice we capture the building and its surrounding context because we needed information for a planning application. The model includes all the relevant surrounding context

The walls we draw are BIM objects and even at this stage we are adding attribute data indicating, for example, that they are existing construction, or if it is evident that a wall is block work or a partition.

At this stage, option studies start to be explored, and doors and windows with attribute data are also added.

Sorsby Medical Practice required external volumetric modelling of the surrounding buildings because we needed to produce information for a planning application. The point cloud is invaluable for this exercise because there is no ambiguity at all. The same clipping, tracing is employed with forms extruded by snapping to the levels in the point cloud

Then windows are indicated in simple form by tracing, projecting and then depressing into the volume of the building. The planning proposals for Sorsby Medical Practice were modelled using this context and visualisations and drawings.

On West Ham Lane, as the design of the model progressed, we used colouration by space type attribute as a way of illustrating to the end users how the spaces would work. In this drawing view of the model the space areas are coloured by their space type differentiating between clinical space, circulation, admin and facilities.

Through detailed design the services engineer provided a sketch scheme which we then worked up into coloured volumes to study the consequences of their proposals. We realised that there were some vertical coordination issues with their roof level drawing showing a connection some 3m away from its arrival 500mm below into the ground floor. We also discovered that they had routed the main air handling supply and return through a room that was intended to remain untouched and so proposed an alternative to this approach.

As already noted the phasing requirements to keep the building in operation while the work to the ground floor took place were complex. To assist the contractor and the management team in understanding what would be demolished and where new partitions would be constructed we produced further diagrammatic visualisations highlighting these on the model.

These supplement the normal drawing package in a way that makes the model more tangible. We also provided visualisations of the different types of partitions illustrated by colour to clarify the requirements.

'The point cloud is invaluable for this exercise because there is no ambiguity at all.'

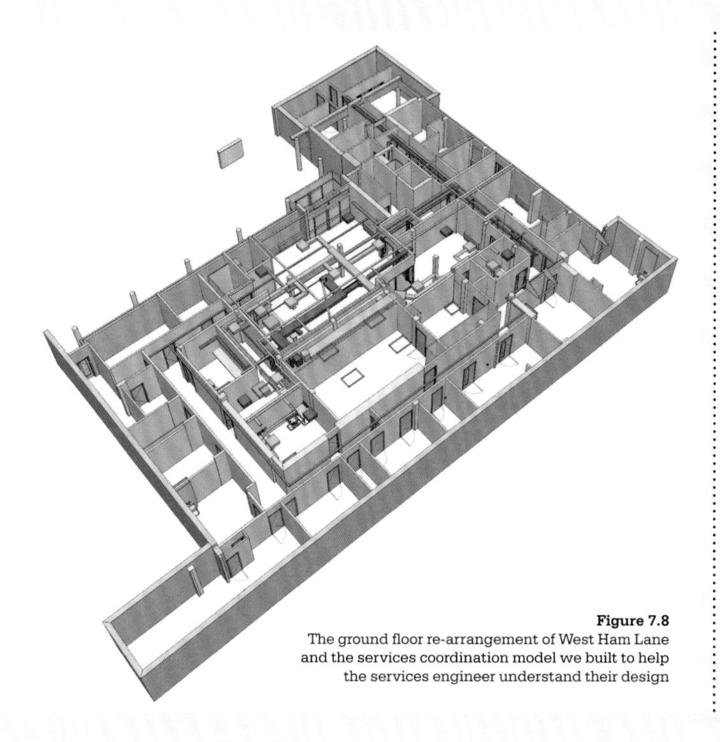

Figure 7.8
The ground floor re-arrangement of West Ham Lane and the services coordination model we built to help the services engineer understand their design

'Through detailed design the services engineer provided a sketch scheme which we then worked up into coloured volumes to study the consequences of their proposals.'

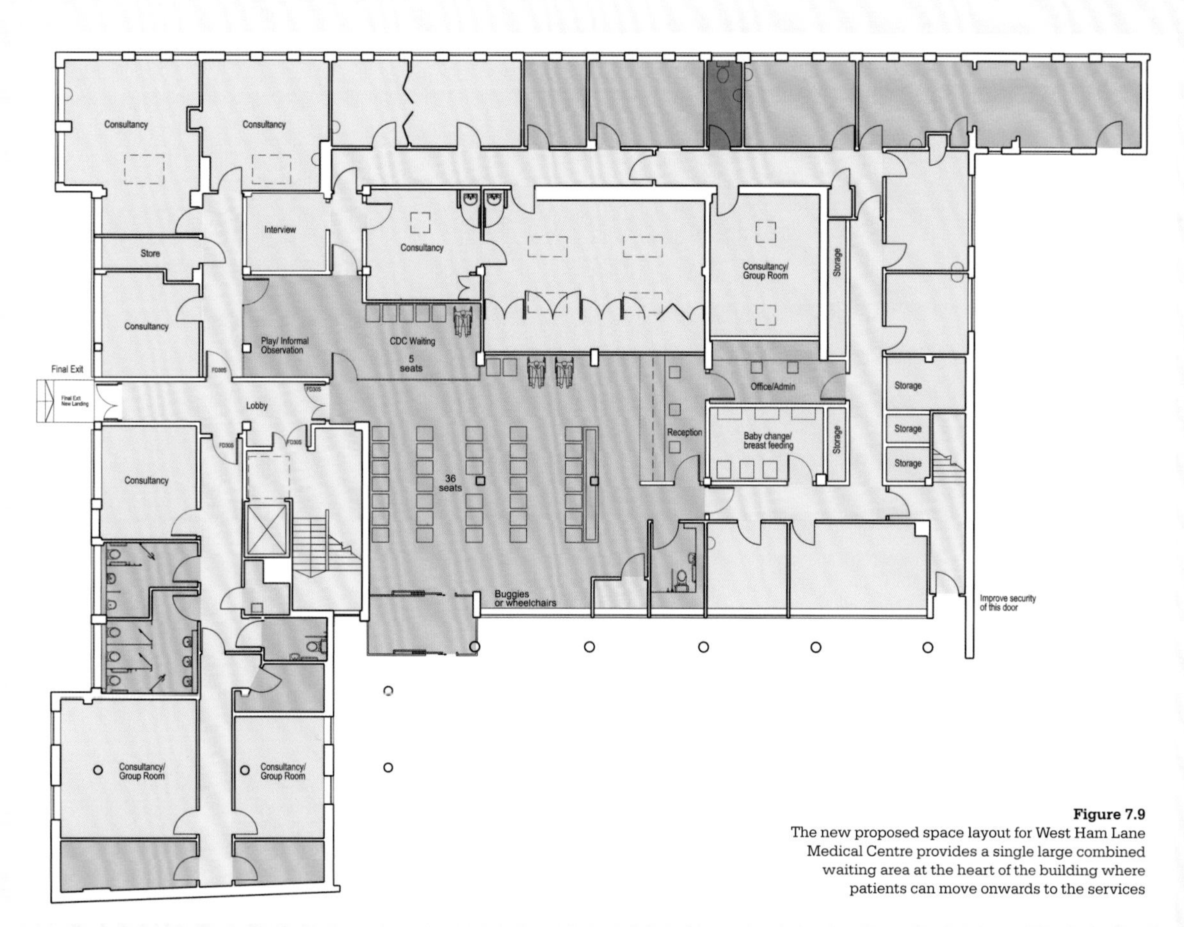

Figure 7.9
The new proposed space layout for West Ham Lane Medical Centre provides a single large combined waiting area at the heart of the building where patients can move onwards to the services

CONCLUSIONS AND FUTURE PLANS

On both West Ham Lane Health Centre and Sorsby Medical Practice for the first time we started to see greater client engagement with the point cloud and existing BIM process. Before these projects we have taken the advantages ourselves. With these we made an active effort to show the technology to others.

The client from the estates department was very impressed with the detailed information that could be seen on the webshare browsing interface and wanted to review having the building captured again after completion. The project manager (who was initially sceptical) is now an advocate of using the technology and the services engineer who still works in 2D has been using the webshare to review data.

The ability to produce accurate existing building information now forms a core part of our business plan, because so few companies are able to and is proving to be a route to securing architectural work.

This, coupled with providing easy visual access to our information in the form of 3D illustrations of the information in the model, makes our practice a much more integrated part of the supply chain and provides a strong incentive to contractors to give us repeat work.

However we're aware that at present we are ahead of the market. In order to stay ahead in this field we've started to look at whether there are parts of the process that we can automate more.

Unfortunately the meshing software that is currently available still comes from industries such as automotive or product design and the costs are higher and so moving forwards with this is reliant on larger projects coming into the office. They are also more focused on meshing smooth continuous curved surfaces such as car body panels or machine parts for which the geometry is reverse engineered. It is just not really tailored to use in buildings.

There are already good reasons to use this software, for example automated meshing of ground models and specialist features in buildings like historic detail. It is simply a question of having the demand for and the volume of this type of work to justify the outlay. Until that point we will continue using less automated workflows.

LESSONS LEARNT

- Being small without legacy systems we are able to use the technology we want and trial new workflows as soon as they are available.
- BIM has become an integral part of the way we get new business and our access to new clients is almost totally related to technology. Whilst there may be a temptation to focus on your skills as a designer this may not differentiate you from other larger, more experienced, practices that do not have your technology skills.
- Understand the programme implications of the technologies you are using. Point cloud surveying is a rapid process but programming the modelling needs to be understood, managed and discussed with other parties who will rely on your information.

BIM FOR PASSIVHAUS DESIGN

Elrond Burrell

Practice:
Architype Ltd

Location:
London and Hereford

Employees:
1 associate,
1 senior architect,
5 architects and
6 architectural assistants

Founded:
London: 1984
Forest of Dean: 1996
Relocated to Hereford: 2006

Hereford director:
Jonathan Hines

Technology:
Vectorworks Architect; Vasari; SketchUp; Revit Architecture; Parallels; Ecotect; Adobe Photoshop; Passive House Planning Package (PHPP)

08

The next two chapters focus on the use of BIM in Passivhaus projects. This case study discusses Architype's use of Revit through some of the more conventional internal BIM workflows, producing conventional architectural outputs. It then gives a detailed dialogue about the practice's customisation of Revit to provide semi-automated numerical outputs from design models to feed into the Passive House Planning Package. This provides a good example of how small practices can build solutions on top of the out of the box product.

THE PRACTICE

We pride ourselves at Architype on being a design-led practice that works across many sectors, with the common thread that we always have social and environmental concerns at the heart of our work.

Our early projects were mainly community and housing projects based around the Water Segal self-build method and innovative use of timber. From this starting point we expanded to deliver projects in the health, workplace, community, public and private institution and education sectors.

In recent years we have had a notable presence in the education sector, particularly for children's centres, primary schools and special needs schools, whilst also gaining recognition for our strength in sustainability and low carbon design. Over the years we have also developed a deep involvement in post-occupancy evaluation and the wider soft landings process.

In 2009 we embarked on delivering three primary schools to the Passivhaus Standard and since then we have progressively moved towards the majority of our work being delivered to meet the Passivhaus Standard and to include soft landings. Currently our portfolio encompasses housing developments, offices, schools, university facilities, an archive and records centre, churches, a hospice, a national park visitor centre, industrial facilities and private houses.

Architype have been a Macintosh-based practice from the very beginning and traditionally used Vectorworks for 2D CAD work. In the early years some ambitious staff experimented with full 3D modelling but found that the hardware and software available at the time wasn't fully capable of supporting such a process.

In 2006 when I joined the office, I brought with me several years' experience of working exclusively in Revit from conception through to completion on projects. At that time Architype was using a mixture of Vectorworks for 2D work along with SketchUp and Ecotect for 3D work and Photoshop for retouching visualisations.

I could see from an outsider's perspective the inherent disadvantages of switching between 2D and 3D in different software packages and by different people – primarily in coordination issues and work duplication. I put forward a case to experiment with using BIM on a project and proposed that Revit be trialled to establish if it would be a feasible undertaking.

Our trial was successful (on our largest project at the time!), even in the face of numerous cultural and technology challenges, not least that Revit only worked on Windows and so we began a process of implementing BIM across the office. We assessed each project to establish if BIM was needed or advantageous and if the staff resourcing would support the use of BIM.

In parallel we continued with staff training and development in using the software and understanding the BIM processes. As we started designing to the Passivhaus Standard we also started using the required Passive House Planning Package and exploring ways of moving useful data between our different software packages. Now our technology workflow involves Vectorworks, Sketchup, Ecotect, Revit, Vasari, Photoshop and the Passive House Planning Package.

THE CHALLENGES OF USING BIM

We have enjoyed considerable benefits from our exploration and use of BIM, however we have also experienced a number of challenges along the way.

There has been a notable learning curve for some people. The change has been particularly challenging for people with extensive 2D CAD skills and expectations that have needed to be unlearned to a certain degree. People in this position are suddenly beginners again, except they know what they want to achieve, just not immediately how to do so.

At the early stages of our BIM adoption I would often hear 'Why can't I do 'X' that I can do so easily in 2D CAD?' and 'I could do this so much quicker in 2D CAD.' These issues took time to overcome as skills increased, mindsets shifted and the benefits became apparent. The flip side is that, once people facing this particular challenge became fluent in BIM, they soon started championing BIM on their projects and didn't want to go back to 2D CAD.

We have also incurred increased hardware and software costs. We had to progressively upgrade our computers, perhaps more so than if we continued to use 2D CAD. As mentioned earlier we are a Macintosh-based practice and remain committed to this.

Unfortunately the majority of BIM software is Windows-based, which limited our choices if we wanted our BIM software to run native on the Macintosh operating system. The alternative option, which is what we have followed, is to run a virtualisation of Microsoft Windows on the Macintosh operating system and then run any Windows-based software in the virtual operating system environment. This means that in addition to the cost of the BIM software, which in itself is considerably more expensive than 2D CAD software, we had the added costs of the virtualisation software, Parallels and the Microsoft Windows operating system for each workstation. It would have been less expensive to go with Macintosh-based BIM software, but we chose the tools that suited our needs and experience at the time.

This set-up also means that the software has more demanding hardware requirements than running BIM software native on Macintosh OS would or 2D CAD software did. However, we have aimed to progressively upgrade our computers anyway so we have just ensured that the hardware performance specifications are suitable for our set-up as we did so. Along the way we also transitioned to iMacs rather than MacPros, which reduced hardware costs slightly and removed the need to purchase monitors separately.

There has also been resistance to change from some staff as not everyone is enthusiastic about changing working methods and learning new skills. Architecture is arguably a tough profession to learn and many people feel there is enough ongoing learning required without the additional challenge of learning new software. Alongside this, not everyone is interested in the broader picture and some people would rather focus on the immediate task at hand, which is often 'producing a drawing'. BIM pushes people to think beyond this and consider the whole package of information required for a design and how it is best captured and communicated. We have managed this in particular by pushing people forward but aiming not to push them too far beyond what they are comfortable with. We allow and encourage people to use whatever software they are comfortable with and is suitable for the stage of the project.

'The change has been particularly challenging for people with extensive 2D CAD skills and expectations that have needed to be unlearned to a certain degree. People in this position are suddenly beginners again, except they know what they want to achieve, just not immediately how to do so.'

THE PROJECT

▶ BUSHBURY HILLS PRIMARY SCHOOL

The brief was for a standard one-form entry primary school (210 children), plus a 30-place nursery and facilities for a local 'Multi Agency Support Team'. A critical requirement was to engage in a thorough consultation process in order to develop a solution that met the educational needs and expressed the ethos of the school in its efforts to provide high-quality education for children from the tough estates surrounding the school. During early consultations it was agreed that the environmental target would be set at achieving Passivhaus, the most rigorous energy standard in the world, so long as the standard available budget and tight timescale were not increased. As we had developed a very positive and collaborative relationship with Wolverhampton City Council over several projects they could see that it was a natural step forward from the previous work we had completed with them.

To achieve the rigorous technical demands of Passivhaus within a standard budget required that Passivhaus be integrated into the design from first principles, informing and influencing every decision about form, design and detailing, together with a relentless focus on simplifying and optimising the design. Simplicity plays a vital role as complexity poses considerable risk to the cost, constructability and quality and the completed building must match the modelled details to pass the rigorous quality assurance requirements of Passivhaus. Under these circumstances and with the skills of our internal team on the project, it was a natural step for us to choose to progress the project with BIM.

The timetable and programme were tight to meet funding deadlines: design team appointment was made in December 2009, construction commenced in September 2010 and handover achieved in October 2011. The budget was a standard allocation within the Primary Capital Programme, which gave a total base construction budget of £4,061,600 for a building of 1,900m².

Figure 8.1
The south elevation of Bushbury Hills Primary School

Figure 8.2
The central hub space with highly sustainable materials and finishes – timber, lino, rubber, mineral paints and organic stains

Figure 8.3
External cladding is a combination of locally sourced brick and UK grown timber cladding, with standing seam metal roofs

Figure 8.4
The main staircase constructed from sustainable timber and plywood and finished with organic stains

HOW WE USED BIM ON THIS PROJECT

There were three key aspects of BIM on this project: the geometry model, the energy model and the flow of data between the two.In terms of the geometric model, we had been encouraging the use of BIM with our design team partners for several years already and so the Structural Engineer joined us on the BIM journey. However, the M&E Engineer still wasn't ready to do so. This meant we could carry out full 3D coordination between the architecture and the structure using a federated model. Because we were both using Revit, this was straightforward without the need to translate files and we used shared coordinates native within Revit.

We used the geometric model for all of our 3D views, General Arrangement plan, section and elevation drawings, internal elevation drawings and schedules. The model was also used to set out and coordinate all the details, but the majority of construction details were still drawn in 2D CAD. This was due to our partnering relationship with WCC where they were producing the initial construction details in-house and weren't set up to use BIM. It was expedient for us to develop the details further in 2D CAD rather than recreate them linked to the model within Revit and this suited the programmatic pressures on the project.

Another application for BIM was in checking coordination and quality control of subcontractor drawings. We imported the 2D CAD files we were sent into Revit and compared them to our model, treating the geometry model as 'the one source of truth' for setting out and coordination.

As this was our first project testing out and developing how we use BIM for Passivhaus, a lot of our workflow was trial and error. We made every effort to use the geometric model in ways that are simple, useful and productive to ensure maximum benefit to the Passivhaus design process. Over the course of this project and subsequent projects we have refined our process into two categories generally: visual interrogation of the model and generating accurate live numerical data.

Visual interrogation is one of the obvious uses in any BIM project because you can see much more in a 3D geometric model than in a 2D drawing set. In terms of translating this to a useful process for Passivhaus we focused on three particular areas:

- Checking the thermal envelope continuity by cutting sections, plans and details through the 3D geometric model and moving the cut-plane in real time to check the continuity of the thermal envelope and identify any areas that need further attention.
- Checking and understanding the heat loss envelope in 3D. It is important to understand and minimise areas that the building will lose heat through; even the most elegant design is best understood in 3D to ensure soffits, protrusions, cantilevers, etc., aren't missed.
- Interdisciplinary coordination and integration. It is often where the architecture and structure and services are not coordinated, particularly in 3D, that unexpected heat loss, energy consumption, CO_2 emissions, potential comfort issues and construction waste or repeat work occurs.

While these visual checks allowed us to find and highlight issues, they did not solve the issues in themselves. There are many functions of the Passivhaus design process that currently cannot be carried out within Revit (or other BIM authoring software). So having found an issue we then needed to work on the design further to resolve it or we needed to remodel the issue in the appropriate software to solve the issue. (for example, thermal bridge modelling software). Once the issue was resolved we then went back to the Revit model to update it.

Figure 8.5
The hall is one area where 3D coordination of the architecture and structure in the federated model was of critical importance as the visible structure is part of the aesthetic

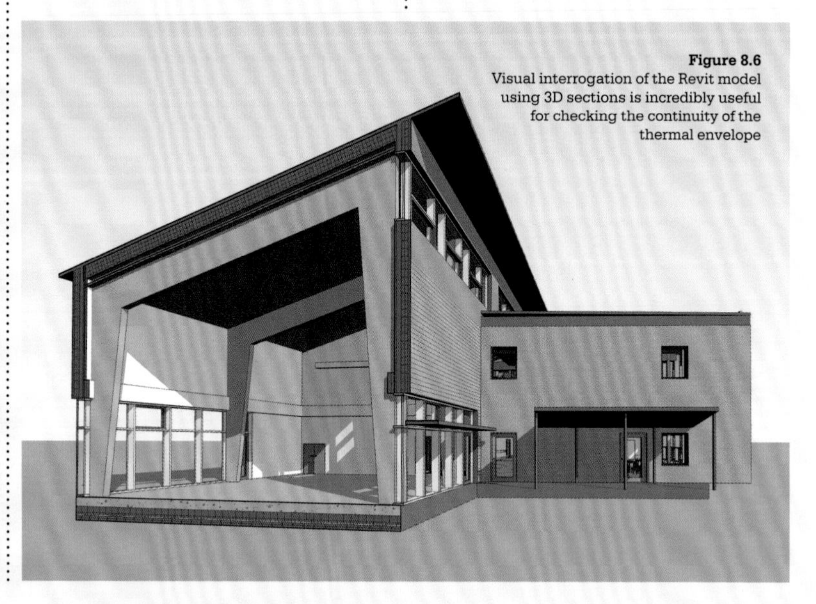

Figure 8.6
Visual interrogation of the Revit model using 3D sections is incredibly useful for checking the continuity of the thermal envelope

Generating Accurate Live Data started out as simply using the geometric model to schedule information as is typical on all BIM projects, however, we focused on what we needed to enter into the Passive House Planning Package (PHPP). This is essentially a very detailed spreadsheet that is used for the comfort and energy modelling required for Passivhaus. Arguably PHPP is a building information model – although it is not a visual / geometric / production information model as most people tend to think of BIM, it is a data-driven comfort and energy model based on rigorous building physics.

We exported data from Revit schedules to a spreadsheet format, manually checked the data and then entered it manually into PHPP (or provided it to a consultant for the same purpose in some instances). As PHPP is used continually during the design process to iterate the comfort and energy performance to ensure required benchmarks are met, getting accurate data out of the geometric model at any point during the design process meant that it was possible to iterate swiftly and accurately, rather than each iteration taking a lot of time and requiring time-consuming and potentially inaccurate manual take-offs from the drawings.

Once we saw how much benefit this provided we looked at ways to make the same data even more useful and ways to generate additional useful data. We found that we could do this by including some factors and calculations in schedules within the geometric model. This had the added advantage of eliminating a further stage of manual effort. For example, floor areas are easy to schedule in the geometric model and previously we had exported these out for manual adjustment to the right format for PHPP. We found we could build factors into the schedule in the geometric model to get all the data in the right format before it was exported. In the same way we were able to use calculations to generate further useful data. The data we specifically focused on was:

- Treated floor areas (TFA), which measures the heated inhabited spaces of the building and is required in PHPP. It also facilitates like for like comparison of performance between different buildings.
- Ventilation volumes, which is a measure of the inhabited / used air volume in the space. Indoor air quality is an important aspect of Passivhaus and accurate calculation of the volume of air that requires ventilating is critical for this.
- Internal building volumes, which are necessary for calculating how airtight the building is when a blower door test is carried out.
- Opening areas (typically windows, ventilation grilles and doors), which are critical for the overall energy performance of the building.

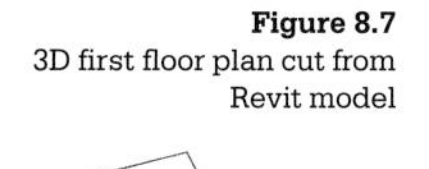
Figure 8.7
3D first floor plan cut from Revit model

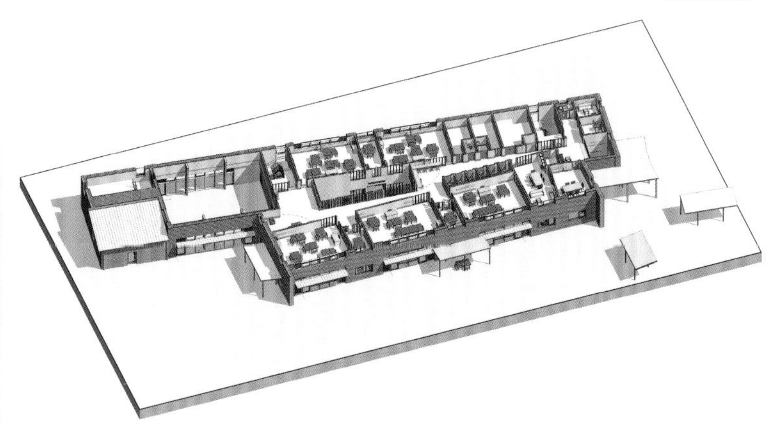

All of these methods involve data derived from schedules in the Revit model that are manually exported to a spreadsheet format and then manually entered into the PHPP spreadsheet. We looked at ways to automate the data transfer process and considered some commercially available Revit add-ons. However, we decided not to automate the process any further as we wanted specific steps in the process that allowed for professional judgement and a quality assurance overview. It is an ongoing consideration though, especially since another BIM authoring software tool, ArchiCAD now has built-in functionality to export data directly to PHPP.

In addition to these more straightforward methods of generating data, on later projects we also developed a method of generating heat loss areas, which are critical for the overall energy performance of the building and need entering into PHPP. However, for this data we need to carry out further modelling as the geometric model used for visualisations and production information in BIM authoring software does not automatically have the right areas. Even so, it is still a more efficient method of generating the data accurately than manually measuring off the 2D drawings with the added advantage of being swift to update at any stage during the design. The methodology we developed has since proved so useful that we also use it on projects where we are Passivhaus consultants and don't actually need to develop a geometric model ourselves.

We use PHPP as a proactive design tool throughout the design process, iterating and optimising as the design develops. The methods described above all allow for rapid exporting of current ('live') accurate data, which is then manually entered into PHPP. At the same time, options and variations are explored in PHPP independent of the geometric model and also as a parallel check when options are being explored in the geometric model. PHPP is an open and parametric model so changes and tweaks to explore options for one aspect of the design (e.g. glazing sizes, shading dimensions, storey heights, building orientation etc). result in instant updates throughout the model including the reported energy/comfort results. This is an important aspect of our design process and one of the keys to our success in delivering award-winning architecture with radical low-energy performance at no additional cost. Too often energy performance is used as a check and measure process once the building is designed, or only at key stages, rather than an integral part of the design process. Or even worse it is a tick-box exercise carried only for rating or compliance purposes.

Data flow between the geometric model and the energy model is clearly critical in this process. Often people assume that the best solution is to automate data transfer as much as possible, but our approach is quite the opposite. As already discussed the process of generating the data in Revit is already fairly automated, however, we treat the data flow out of the geometric model as an opportunity to both carry out quality control checks and as an opportunity to exercise professional judgement. Our experience in designing sustainable architecture and our judgement plays just as important a role in the design process as the numbers and graphics from BIM! So we specifically export data out of Revit and manually parse it and re-enter what we need to into PHPP to suit our current design process needs. We are aware that this does introduce an opportunity for human error.

However, we feel that this risk is considerably lower than the risk of errors arising from assuming that an automated process is both accurate and precise. Computer-generated numbers, charts and graphics easily seduce too many people who don't actually understand what they mean, if indeed they do actually mean anything useful!

WHAT WE LEARNED ON THIS PROJECT

The project was delivered successfully and certified to the Passivhaus Standard and has since won several awards. Using BIM in the process helped us in a number of ways. It was particularly noticeable that tasks and activities relating to the Passivhaus design process were carried out faster, with more accuracy and more reliably. For example, generating volumes directly from the model saved a person a whole day of manual measuring and similarly with TFA. This meant that we were able to iterate and explore more design options with confidence knowing the data was there and available for use at any time should it be needed.

We also noticed that even though the details were mostly in 2D CAD as mentioned earlier, they were still considerably easier to coordinate and ensure they were accurate. We were able to use underlays exported from the geometric model to check and coordinate everything, whereas on other jobs checks had to be carried out manually and coordination carried out across multiple drawings, all of which is very time-consuming. Although this seems like a very basic benefit on one level, the knock-on benefit, which is key for us, is that we have more time available to devote to research and innovation so we are able to keep innovative projects within standard budgets and time constraints.

Bushbury Hill Primary School
Bushbury Hill, Wolverhampton, West Midlands
Ground Floor Plan

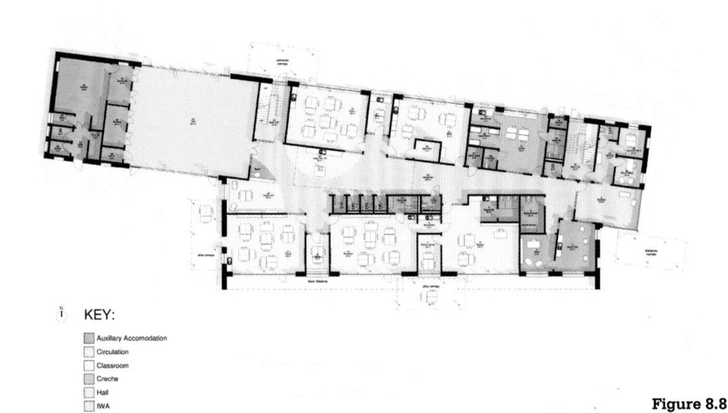

Figure 8.8
Ground floor plan

Bushbury Hill Primary School
Bushbury Hill, Wolverhampton, West Midlands
First Floor Plan

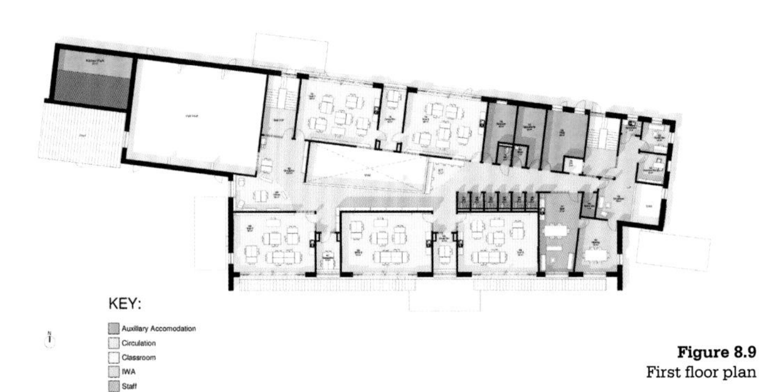

Figure 8.9
First floor plan

As this was our first project to use BIM specifically for Passivhaus it gave us many opportunities to learn and develop our methodology. We were quite focused in our approach; nevertheless, we didn't know for certain what aspects of BIM we would end up putting to good use for Passivhaus when we started the project. The techniques set out above were tried for the first time on this project and have been considerably refined over subsequent projects.

As part of our learning and refining process arising from this project, we set up a specific project template with drawings, views, schedules and parameters that apply to Passivhaus built in from the start. This means that when a new project is set up the information we want for the Passivhaus design process is generated automatically while the geometric model is being built without any extra effort needed. Over the course of several more projects we have gradually developed and refined this further, learning from what works best and what is most useful.

Having the Structural Engineer also working in BIM was a considerable advantage. The fact that the M&E Engineer was not at that point using BIM made the difference all the more noticeable. Happily they have since started their own 'BIM journey' and on a more recent project the same design team was able to fully coordinate the architecture, structure and M&E in 3D. The advantages of this working method were immediately apparent when modelling the plant room. Instead of ending up with an ill-conceived and congested layout, the M&E Engineer's input into the model resulted in a well-coordinated design, with plant equipment placed in an accessible and well thought out position.

CONCLUSIONS AND FUTURE PLANS

Working in BIM has changed our workflows a little, but we have been conscious to use BIM to our advantage. We haven't let the change in software or processes dictate how we design. It has certainly given us considerable advantages and when Wolverhampton City Council commissioned us to design another Passivhaus school on a tighter budget and further reduced time frame we knew with confidence that we could deliver a Passivhaus certified school within the budget and time constraints. Not only this, we also knew we could still afford to invest considerable time in client and user consultation, something we feel is absolutely crucial, knowing that our BIM design process had production efficiencies built in that would allow us to meet the demanding time constraints.

We are currently in the process of implementing BIM across the whole office and at the time of writing approximately 80% of our work includes BIM. We have employed a 'just in time' approach to training. We train relevant staff as they begin on a project, right when it is needed, rather than taking a broad-brush approach and providing generic training for all staff. In the same vein we aim to have as much training carried out in a live project environment so lessons are real and immediately applicable. When we first started using BIM we did undertake considerable classroon-based training but found that the lessons were quickly forgotten and in many cases had a high proportion of redundant content.

Partly as a result of our capabilities with BIM the office has recently expanded quite considerably, growing to 26 people at the time of writing. We initially looked for staff that could use Revit and had knowledge of the BIM process but found that it wasn't a deciding factor in the end. Most people pick up the new skills and knowledge needed when immersed in a project environment quickly and the process encourages peer-to-peer interaction and learning.

We continue to encourage consultants we work with to engage with BIM if they haven't already, as we feel there are so many clear advantages when working together in a BIM process.

We are actively improving and refining our BIM workflow for Passivhaus design. Now that we have an established project template that provides a lot of what is needed from the outset of a project, we are testing it on different projects and with different in-house Certified Passivhaus Designers. At the start of our BIM and Passivhaus journey we specifically aimed to maximise the value we could get for the Passivhaus process without changing the design and production information modelling process. We have now reached the point where we are starting to adjust our modelling processes to increase the value that BIM can provide to the Passivhaus process. For example, previously we modelled windows for what suited drawings and schedules based mainly on the visual appearance and the procurement packages (production information). In this approach a single window component might include several glazed panels and mullions / transoms. However, for Passivhaus each window panel needs specific performance and geometric data and additional information about the shading and adjacency conditions. Providing this in the BIM process involves extra modelling work, either in creating bespoke BIM components or in adding bespoke parameters to pre-existing BIM components. We don't have a conclusion to this particular situation yet as we are still exploring what works best.

There are also a number of gaps in the useful information for Passivhaus that BIM software and Revit in particular, makes accessible. For example, component or element orientation can't be scheduled in the model. Because the model can be located and orientated correctly in the 'real-world' location and orientation, clearly the information is there in the model. There are a number of commercially available Revit add-ons that do make this information available (along with other equally inaccessible but potentially useful information) and we may consider investing in these tools at some point to extend the capabilities of Revit.

Meanwhile we have observed that ArchiCAD, one of the competitors to Revit for architectural BIM authoring, has invested considerably in improving integration with Passivhaus design and PHPP data requirements. This might be due to ArchiCAD's strong European base where Passivhaus is well established in comparison to Revit's stronger US base where Passivhaus is still relatively new. Hopefully this kind of healthy competition will result in all software providers continuing to improve their offering.

LESSONS LEARNT

- Look for ways to automate your workflow as much as you can but not too much. Professional judgement and subjective quality control based on experience and knowledge still have important roles to play.
- Use BIM to filter the available model data for what is actual useful information. BIM software has the potential to provide a huge amount of data from the model but only some of it is actually useful. Having the right data available at the right time is more important than having a vast quantity of data, particularly when designing to a rigorous energy standard such as Passivhaus.
- It is important to tailor technical support to an individual's needs so they can overcome the particular challenges they face.

BUILDING PASSIVHAUS WITH BIM

James Anwyl

Practice:
Eurobuild

Director:
James Anwyl

Location:
Sussex

Employees:
1 director,
1 architect,
1 architectural assistant,
1 design engineer and 1 finance officer

Founded:
2006

Technology:
ArchiCAD;
EcoDesigner STAR;
MEP Modeler;
BIMx (Graphisoft);
TAS (EDSL);
PHPP (Passive House Planning Package).

09

In the second case study on the use of BIM in Passivhaus projects Eurobuild talk about the effect that BIM has had on the construction phase and beyond. Like Architype the practice uses a wide range of software to produce its information. This chapter discusses sequencing and coordination from the model and then elaborates on how the model can contribute to the post-construction phase of a project, with a review of how it was used for health and safety, O&M Manuals and the BREAAM User Guide.

'We began by using the online tutorials and within two weeks we were up and running and more importantly knew 80% of the functionality of ArchiCAD.'

THE PRACTICE

Established in 2006, Eurobuild is an RIBA Chartered practice based in Sussex. We have a core team encompassing a practice manager, architect, architectural assistant, design engineer and finance officer. We also work closely with a wider team of associated engineers and further specialists.

Eurobuild specialise in the design of Passivhaus and low-energy buildings. Passivhaus typically achieves more than 80% reduction in energy consumption over existing buildings, resulting in significant reductions in operating costs and maintenance.

All our projects since 2007 have been designed using BIM. We are advisors and beta-testers for Graphisoft. Eurobuild was one of six architectural practices advising the government on BIM implementation through the Technologies' Alliance and have advised a number of main contractors on how to make effective use of BIM. We began using BIM because our existing software licence was coming to an end and we were researching other potential software.

With the ability to work faster and with a reliably coordinated output it soon became apparent that there was no turning back from working in 3D. We began by using the online tutorials and within two weeks we were up and running and more importantly knew 80% of the functionality of ArchiCAD.

The programs we use include:

- ArchiCAD 17
- EcoDesigner STAR (Graphisoft) – for early stage energy evaluation
- MEP Modeler (Graphisoft) – for integrated services design
- BIMx (Graphisoft) – for presentation and client walk-round of entire model
- TAS (EDSL) – thermal analysis simulation
- PHPP (Passive House Planning Package).

BIM is not just 3D design, nor is BIM just software. BIM is at the core a process and provides a way of rehearsing the construction of a building virtually, element by element.

Each element can be attributed with useful information like cost, size and weight, material type that can also be located in the virtual building spatially by storey, zone or room.

The sustainability agenda is driving architects to develop designs that look beautiful and feel great but also reduce energy consumption, carbon emissions and operational costs.

Development of a project model using BIM allows us to design to a high level of complexity and the software can help to predict the energy use of the building.

In the early days, we struggled a little with how to classify an object, which layer to put it in and to what level of detail we should be working. Our protocols are becoming more established with each job and they also evolve each time so tracking the changes is crucial.

It was important to define a template so that we all work in the same way, but we are still wrestling with what level of detail to provide – because it's becoming very easy to work at high levels of detail, it's very tempting to provide it. Then you have to manage client expectations because they get used to you over-delivering!

THE PROJECT

Completed in 2010, The Rural Regeneration Centre at Hadlow College in Kent was the first certified Passivhaus educational building in the UK and is located on the college's fully operational dairy farm.

Eurobuild inherited the project from another architectural practice that failed to get planning approval. The main purpose of the building is to enable seminar-based teaching, with a staff office and meeting space alongside an exhibition area.

Technically a conversion, the building is 90% new-build but uses some block-built elements of a redundant cow shed for the plant room and wet working demonstration area. The project was shortlisted for BUILDING's Sustainable Project of the Year 2010 and won Environmental Project of the Year 2011 in the Construction Computing Awards for innovative and comprehensive use of BIM.

We now have two years of data from the energy monitoring system and this has been analysed to assess the performance. We compared the results with a survey of 834 schools built in the last ten years and compared to the mean averaged result from the survey (133 kWh/m^2/yr), Hadlow's RRC is using 80% less energy for heating and hot water and 70% less electricity. When compared to the best performing school in the survey, Hadlow's RRC uses about 50% less heating energy and more than 50% less electricity.

Possibly more importantly, Hadlow's RRC cost 15% less to build than a standard school at the time – £1742/m^2 including all fees and landscaping, of which £1484/m^2 were construction costs. We could not have achieved this value for money without ArchiCAD.

▶ IN AREAS OF COMPLEXITY

This was where some parts of the old building needed to be added to, before joining to the new parts of the building. The build sequence was simulated for the contractor and the QS to make clear what was happening, so that costs could evolve accurately. This was a precursor to the renovation tool developed by ArchiCAD, which we use a lot now in refurbishments.

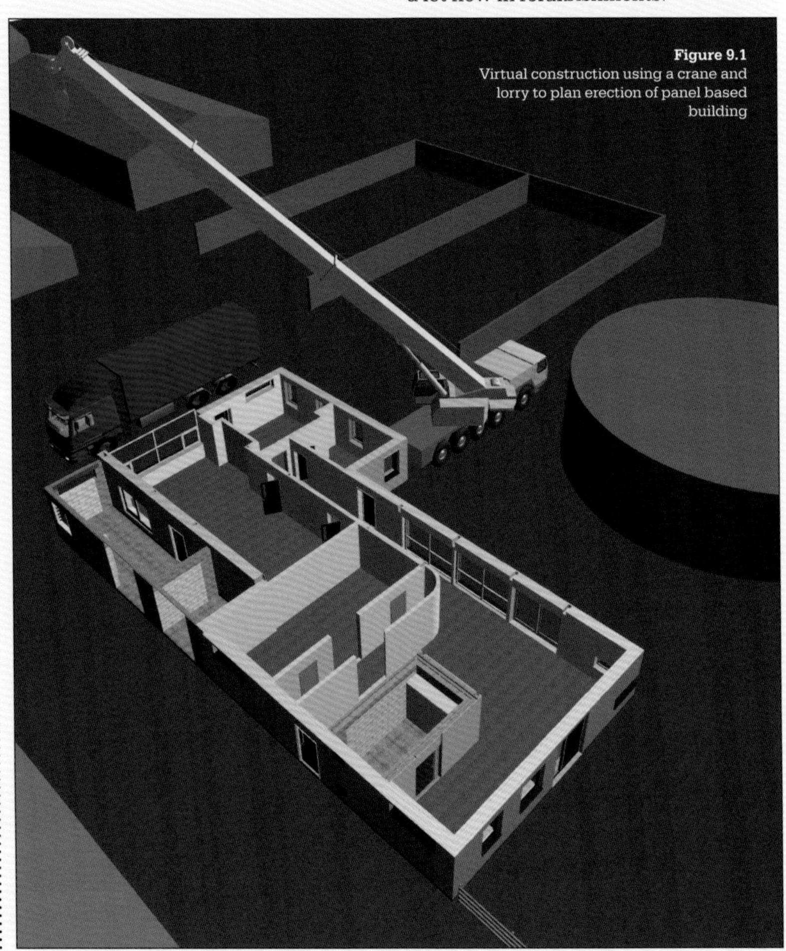

Figure 9.1
Virtual construction using a crane and lorry to plan erection of panel based building

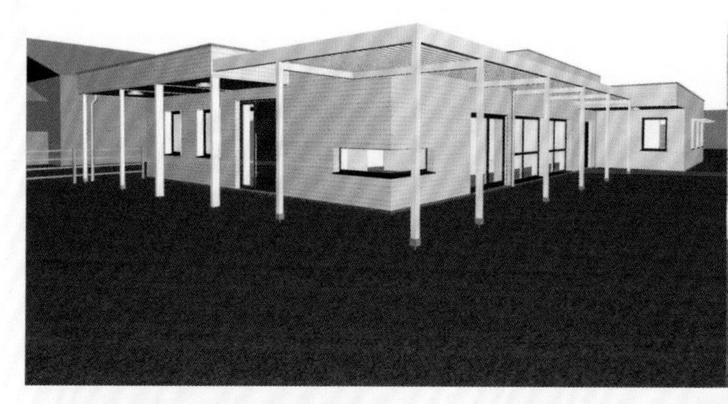

Figure 9.2
In BIM and as built: ArchiCAD's accurate representation pre-construction

▶ METHODS STATEMENTS

The method statement for the crane schedule lift of components was in the form of a visual plan. The project CDM coordinator stated that this communicated more clearly than any written statement could. It also allowed for maintained access zones, for example it defined where the lorry and the crane would be sited to avoid clashes despite a high quantity of material on a constrained site.

▶ VALUE AND AFTERCARE

In terms of aftercare, a website was built to be used by the client team and hosted by Eurobuild that combined all the information required for the Health and Safety, O&M (Operation and Maintenance) and the BREEAM User Guide. This was more cost-effective than creating the standard lever arch folders and resulted in a more accessible and updateable facility.

The cost associated with assembling these documents in 'analogue' folder format equated to the same amount as creating a website for all the data. Using hyperlinks to reduce document volume enables updates, etc. with no chance of documents getting lost.

▶ CHALLENGES AND OUTCOMES

In terms of challenges the building had a full-height, north-facing window to provide light to the main seminar room. This feature became an issue when running it through the Passive House Planning Package (PHPP) and required increased performance in other areas of the building to counteract the heat loss.

This included designing in solar gain to the south and south-west elevations and from skylights, as well as upping the wall insulation from the usual 300mm to 400mm. These performance improvements were modelled in to ArchiCAD to assess their visual impact and thermal performance.

The seminar room windows are time and temperature operated to enable free cooling at night with effective cross-ventilation. Individual window pergolas and a colonnade across the south façade prevent overheating in summer – all solar modelled in 3D ArchiCAD, checked with TAS software and backed up by the PHPP analysis.

Constructed of super-insulated closed panels, the structure was planned and assembled in just three days. The build process was communicated to the site operators with such clarity that the result was seamless assembly of the 350m^2 building.

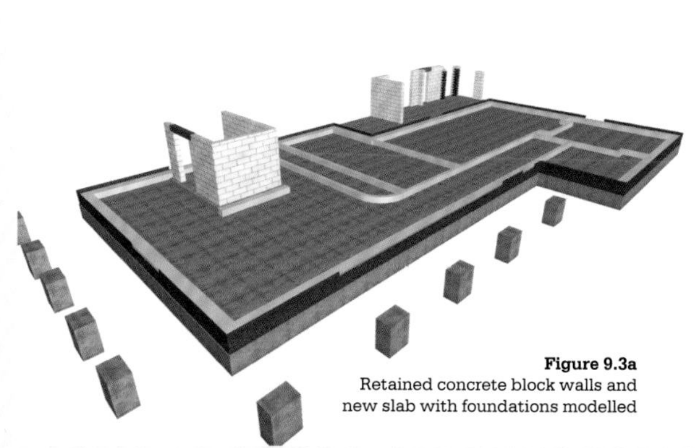

Figure 9.3a
Retained concrete block walls and new slab with foundations modelled

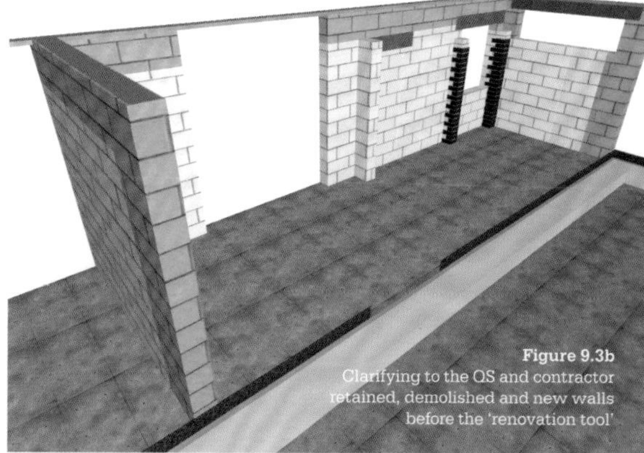

Figure 9.3b
Clarifying to the QS and contractor retained, demolished and new walls before the 'renovation tool'

The structure was airtight to a very high standard of 0.34 h-1 in under ten days overall. Without ArchiCAD, this level of prefabrication and subsequent time savings would not have been possible. Using BIM throughout the project, from planning to commissioning, has led us to introduce an in-house BIM standard to ensure consistent output across future projects.

We carried out a detailed analysis of the benefits of MEP modelling the ventilation at Hadlow. We assessed who benefits from this exercise and why. For instance, main contractor benefits from there being less BWIC, whilst OS can generate more accurate quantities and the sub contractor can prefabricate more, reducing waste and time on-site.

Against each of these benefits we calculated the cost saving and defined whether it was a material or a labour saving.

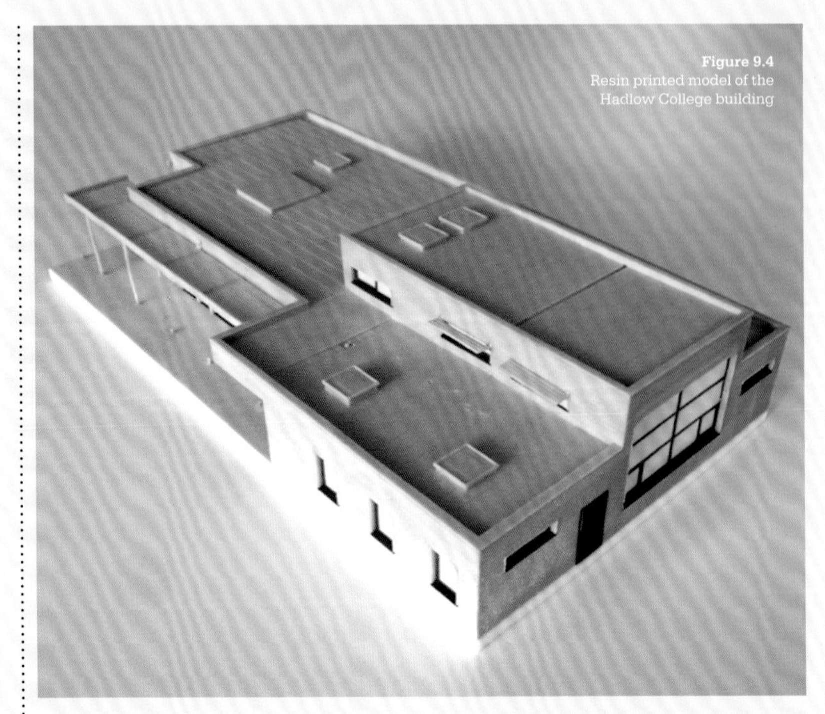

Figure 9.4
Resin printed model of the Hadlow College building

Even with the additional cost (which we had to absorb) of modelling from the mechanical engineering drawings, the net benefit from this single line within the build programme, the net gain was over £3600.

More importantly perhaps, the greatest benefit we see from using BIM is reducing risk, which is a lot harder to monetise or assess from a financial perspective.

CONCLUSIONS AND FUTURE PLANS

The efficiency of working in 3D has revolutionised our process and project outputs. With minor changes there is less checking and slick documentation. Major changes are quick and real. It has driven a far greater understanding of buildings.

The risk is also far less for the same effort, which as a result is demonstrating good value for less input. Thus a section that would normally take a day to provide can be done in less than an hour and be guaranteed to be error free – or as good as the model at any rate!

Using BIM in this way has helped us to develop new relationships and increase our profile considerably, although it takes time for this to translate into actual project work.

We are certain that we do now have more weight with our offer at the tipping point of securing projects. We are also now able to do virtual prototyping which we could not before. The Hadlow College resin model was printed directly from an ArchiCAD .stl file at a scale of 1:100. This was the first time that an object has been printed directly from an ArchiCAD file.

In terms of the concrete benefits of the tools we use the big advantage of using EcoDesigner STAR in the design of low-energy buildings is that it can accurately assess energy performance without exporting to another tool, so all the information built into the composite wall build-ups, etc. can actually be used directly.

In addition, it takes 2D details from ArchiCAD and calculates psi-values in order to assess thermal bridges. Openings are recognised and their thermal properties can again be determined in the tool.

We have also started to use customised technologies to suit the way we work. We have now coded a set of GDL scripted parametric objects that correlate precisely to our timber frame factory parameters, adhering to known engineering and transport limits.

This was developed with our in-house knowledge of the panels system we import from Austria. When we build in ArchiCAD using this system, we can schedule out cost, weight, volume and carbon footprint data in real time, allowing us to assess the design against these factors throughout the process.

It has changed our environment because we all work in ArchiCAD 'Teamwork' now on one model (only) in the cloud – something we didn't have on the Hadlow project.

For Hadlow our mechanical and electrical engineers produced 2D data which were imported as layers and 3D information was modelled by our team and confirmed as correct (to be issued) by the consultants. This was of course time intensive, but massively reduced risk and definitely improved overall design and execution.

To give you an on-site example, our Hadlow project ductwork contractors had estimated four days of work on-site, but actually completed their works within 2.5 days. The 3D drawing allowed them to work considerably faster as they were far clearer on the work required.

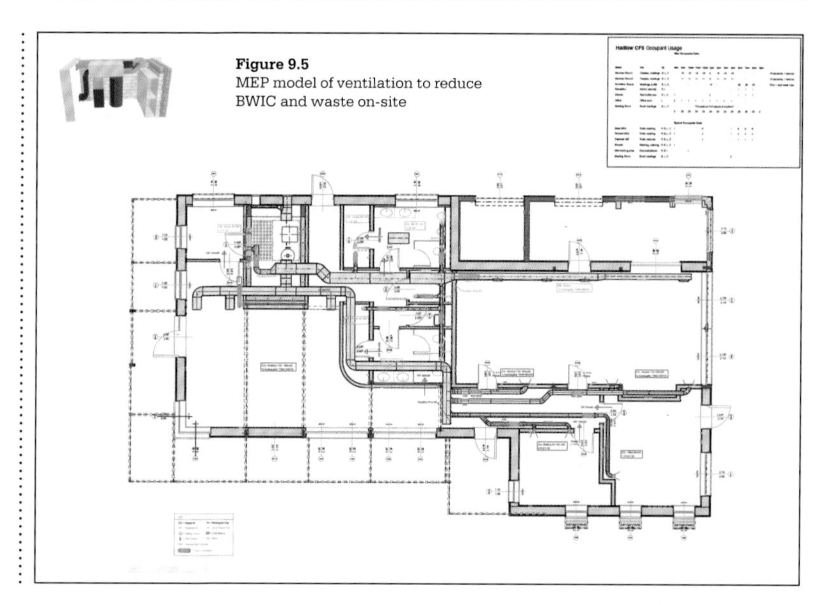

Figure 9.5
MEP model of ventilation to reduce BWIC and waste on-site

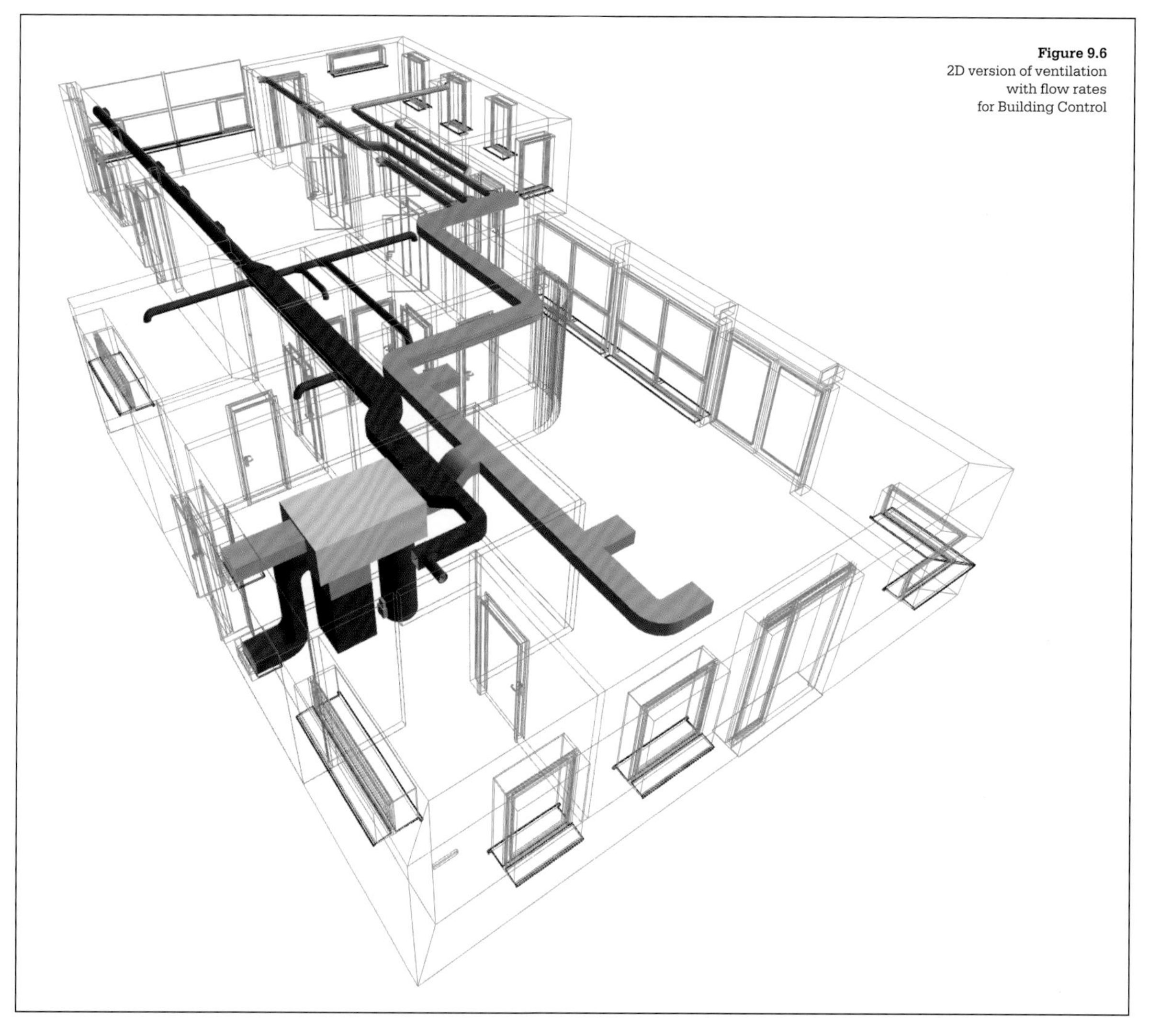

Figure 9.6
2D version of ventilation with flow rates for Building Control

LESSONS LEARNT

- BIM is such a new tool, the breadth and depth of it is vast which allows practices to develop bespoke processes to suit both their and their clients needs.
- You can invest in a single licence for relatively small initial outlay and we would say that you would see the payback very quickly, within a month.
- Go for it!

BIM AND WORKING WITH CONSULTANTS

Nick Allen

Practice:
Metz Architects

Location:
Leicester

Directors:
Nick Allen

Employees:
4 architects, 1 senior technologist / BIM manager, 2 BIM technologists and 1 office manager

Founded:
1998

Technology:
Autodesk Revit Building Design Suite Premium; Autodesk Navisworks; Quarter1; Tekla BIMSight; Solibri Model Viewer; SketchUp

10

This case study focuses on the transition from 'lonely BIM' (maturity Level 1) to working with other consultants (maturity Level 2).

Moving beyond simply taking the internal productivity benefits of BIM to working with others is a very significant step that requires the use of new tools and management processes.

The greater level of awareness of your fellow consultants and their information requirements and timings is key to the success of a project. The case study delves into the detail of making these processes work and the management tools, including a BIM execution plan, that are required to do so successfully.

THE PRACTICE

Metz was formed in 1998 between myself and (then) business partner, an architect from commercial practice. I had been a sole practitioner for ten years, having set up shop in the late 1980's construction boom. In 1998 we had recently won a major national design competition (the redevelopment of Smithills Hall and Estate in Lancashire into a regional visitor attraction) and had secured a stream of work delivering shop fitting production information packages across Europe for the sportswear retailer, Nike.

Previously we were predominantly drawing board based, but in 1998 we dabbled with MiniCAD (now Vectorworks) on Macintosh on a project or two. Nike mandated that we used AutoCAD for their projects so that we could deliver their works packages electronically by the newfangled 'email'. At a stroke then we abandoned MiniCAD and Macs and went over to PC and AutoCAD.

So by demand rather than design, the predominant legacy supplier of software to the practice has been Autodesk.

The economic chill of autumn of 2008 was a seminal moment for Metz. We had just completed the pre-contract design and production work on our largest ever single project; a £38 million college of further education, possibly the last procured before the collapse of the Learning and Skills Council's 'Building Colleges of the Future' capital expenditure programme. The delay and then cancellation of this single programme laid waste to dozens of architectural practices both large and small. This was bad enough, but the decision by the government in 2010 to abandon significant investment in public sector projects wholesale would devastate the profession and push the construction industry into freefall.

At the height of the 'boom', Metz employed 16 staff, had taken a lease on sizeable offices and had ambitious plans for expansion. In November 2008, we surveyed the rows of empty workstations at which, a few months earlier, work colleagues sat, many of them friends.

In 2009 myself and my fellow director made a board decision for fundamental change. On viewing our finances it became clear that big is sometimes far less than beautiful.

We had haemorrhaged resources and therefore money on our 'big break' project, leading to a significant loss.

We had taken on expensive offices and been exploited by staffing agencies. Fees had been undercut for years in spite of the boom. Experience told us that twice as many employees did not translate into twice as much profit (or even any profit sometimes). It was apparent that the practice of architecture, in the traditional sense, was marginal at best. With our major workflow gone and the prospect of a downturn lasting for up to a decade, we either fundamentally changed the way we worked, or went out of business.

We had some luck. We exploited a break clause in our lease and moved to new offices half the size (our workforce was half the size by now after all). We also had a decent project in progress (a remnant from the college building programme without any government funding). Finally, the Autodesk marketing machine had cajoled a couple of our team to a re-seller presentation flogging something called 'Revit' on the promise of a discount if we bought anything. The rest is BIM history (not that we called anything BIM in 2009).

Our team was convinced that implementing Revit (or something similar) would offer us significant productivity improvements as we could generate any plan or section that we cared to, at any location in the building, if we constructed a single comprehensive 3D model.

We could automatically schedule off (doors or windows for example) which would eliminate mistakes if there were revisions, as there would be a single point of maintenance for the information. Better still, if we changed something such as a floor to floor level for example, we could revise potentially hundreds of section and plan drawings in hours instead of days (or even weeks).

At a stroke, we added a tier of delivery efficiency to our technical production and removed a raft of risk. This led to an immediate improvement in profitability.

SMALL AND FLEXIBLE BEATS BIG AND BULKY

Small practice is the ideal springboard into the new age of BIM. We are much better placed to adopt BIM because small practices are more adaptable and flexible than large practices when it comes to managing change. The adoption costs are comparatively low. For the large architectural practice, a boardroom decision on a new venue for the Christmas party is akin to turning around a super-tanker.

Imagine then the logistical and financial dynamic of a fundamental change like adopting BIM now that the tidal wave has breached the horizon. Perhaps one of the biggest assets of running a small practice is to be at the helm of a jet ski. We can make instant decisions and then just do it!

DO YOUR RESEARCH!

All the mainstream BIM authoring tools have their strengths and weaknesses. The origins of Revit are derived from a product called Pro Engineer. It's completely different from AutoCAD so previous experience with Revit won't necessarily mean you're ahead of the curve.

It might be better to start learning a BIM authoring tool that you have no previous experience of rather than relying on an entrenched existing skill set.

Get demonstrations from all the vendors and think carefully about what you expect from the software. Do you just want a 3D modelling tool, or a serious collaborative BIM tool which will integrate with another consultant's software?

Costs for software vary wildly. There are always deals to be done and offers. Get several prices from different resellers. Upgrading from 2D packages is a cheaper option, as is swapping from one vendor to another, as they are all keen for new customers.

Some software products and model viewers are actually free! You can federate and clash detect models in Tekla BIMsight without lashing out thousands on Navisworks for example. You can share and comment on models with your clients for nothing, using some of the mainstream products like Navisworks Freedom.

You should also consider a subscription service rather than a one-off purchase. It seems more expensive but all of the major vendors update their software with better functionality every year and if you skip a couple of releases, you will soon find yourself out of date and could have compatibility issues.

Remember to research the hardware too. Big models are very memory intensive and different BIM authoring tools have different memory requirements so you may find that your existing machines will not cope with your chosen software. Some products will only run on 64 bit machines and you need two big monitors for maximum productivity as well as a decent graphics card.

The decision to adopt BIM was an obvious one although we rather naïvely did not investigate the other options to Revit. We had some spare cash to buy two seats of the software and three days of training for two of our technologists. Our existing hardware could just about cope. In all the initial investment was about £9,000. This may seem significant. If you also add in new hardware costs you are looking at the thick end of £5,000 to £6,000 a seat.

However, imagine you had 50 or 100 staff to train. The costs in cash and time are enormous for larger practices. So much so it's holding many of them back.

Our next decision was how to implement. Some people suggest running a parallel project in 3D alongside a project in 2D (Revit versus AutoCAD for example). The project we had in mind for our first attempt was a good sized one at £16 to £17 million; a quite complex college building containing an auditorium, fly tower, sports centre, music and arts facilities, traditional classrooms and workshops, part refurb but predominantly new-build. Having downsized considerably from our peak, we had neither the available resources nor the spare cash to twin track a project. So rather ambitiously, we decided to go for it – head first into the deep end.

THE PROJECTS

▶ DUDLEY LEARNING QUARTER PHASES 2 AND 3

These two projects form part of a major Further Education campus redevelopment and demonstrate our journey from lonely to collaborative BIM.

▶ PHASE 2, DUDLEY EVOLVE

Dudley Evolve is a purpose-built resource for students over 16 housing learning facilities for business, information technology, travel and tourism, sport, public services, performing arts, hair and beauty, media, graphics, art and design, along with a new stage school for performing arts, dance and music.

Figure 10.1
Dudley College: 'Evolve'

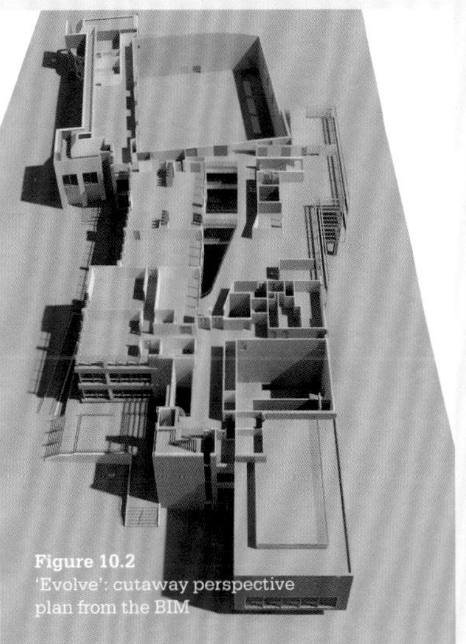

Figure 10.2
'Evolve': cutaway perspective plan from the BIM

Figure 10.3
'Evolve': atrium visual from the BIM

This first project was not really BIM at all. It was 3D modelling with bells on. BIM is about a collaborative working environment with others. This project was a classic example of what we now call 'lonely BIM', which is basically 3D architectural modelling with productivity benefits. In 2009 there were few Structural or Building Services Consultants using BIM collaboratively. Indeed, there were few credible mainstream BIM products for Building Services until 2012. We didn't actually think much about creating a collaborative model. The other consultants we were working with weren't interested and had neither the software nor the inclination amongst the staff to change their skill sets. You need a collaborative model to really tease out some of the fundamental issues in construction regarding miscommunication and misunderstanding between consultants.

This project worked really well for us. True, there is significantly more work to do 'front end' to produce the initial model, but once you have you can greatly step up the productivity. The resource to deliver the tender and production information was significantly reduced to a core team of two technologists working on the model as multiple drawings could be produced at the same instant with related scheduling cross-referenced automatically produced.

There was no requirement for a BIM deliverable from the client group. However, it enabled us to produce 3D views on demand and at an early stage in the design process which meant that

Figure 10.4
Dudley College 'Advance': artists impression (CGI)

the client and stakeholders could visualise the design and spaces straight away. Subsequently they were much more engaged within the design process, which improved decision making. It was only when we actually built the building and moved onto Phase 3 that we realised what the true potential of collaborative BIM was.

After Evolve, we decided our next project would be the full collaborative BIM experience and with the backing of the client, we would force the other consultants to engage.

▶ PHASE 3, DUDLEY ADVANCE

Dudley Advance is a four-storey Further Education building also within the Dudley Learning Quarter designed to deliver advanced engineering and manufacturing skills. At a contract value of £6 million, it contains workshops, laboratories, virtual prototyping facilities, CAD studios and traditional classrooms. Its design reflects the engineering aesthetic. To paraphrase the government's former chief construction advisor Paul Morrell…

'BIM technology could herald the return of architects to the role of 'master builder.' Architects could choose to become 'integrators' of projects 'striving to return to the role that many still romantically profess an ambition for''

This was Metz ambition for our next building.

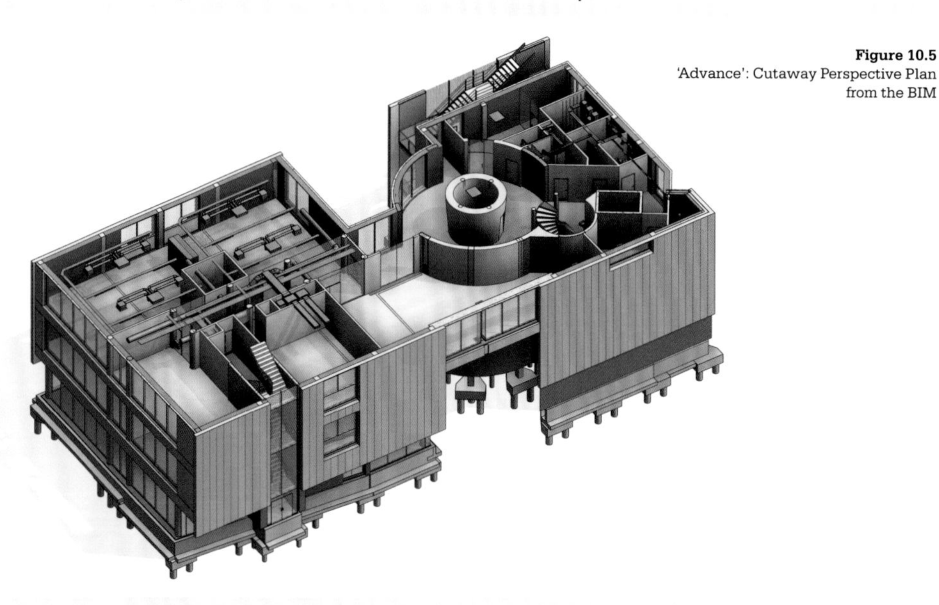

Figure 10.5
'Advance': Cutaway Perspective Plan from the BIM

UNDERSTANDING WHAT EVERYONE DOES

'Coordination of building elements on-site can be a real problem – and this is where BIM can come into its own if used properly. It has the potential to support successful collaboration amongst consultants and contractors.'

Two common issues are:

- The lack of understanding within design teams (and I include architects here) as to what the other consultant members actually design and produce and a lack of proper coordination of those team members' outputs. This has become known as the 'silo mentality'.
- The associated lack (under Design and Build) of pre-contract detailed design information leads to the bizarre situation where the design and selection of large elements of a building and by implication the associated coordination issues, is passed over to subcontractors and fabricators (post-tender) and in particular within the realm of building services.

This invariably leads to significant coordination clashes between building services, structure and architecture. The result is huge cost on-site and compromise on design (often leaving architects wondering why buildings sprout this and that appendage, mystery bulkhead and inconvenient 'boxing in' as their building nears completion).

With BIM, once you understand the concept of reviewing complete buildings proactively in real time as 3D models during design, rather than reactively when the problems arrive on-site, consultant teams can finally understand the implications of their collective design decisions.

Everyone's work is defined visually and is therefore completely transparent to everyone else. Really, once everyone gets out of their silos and joins in, life gets much easier.

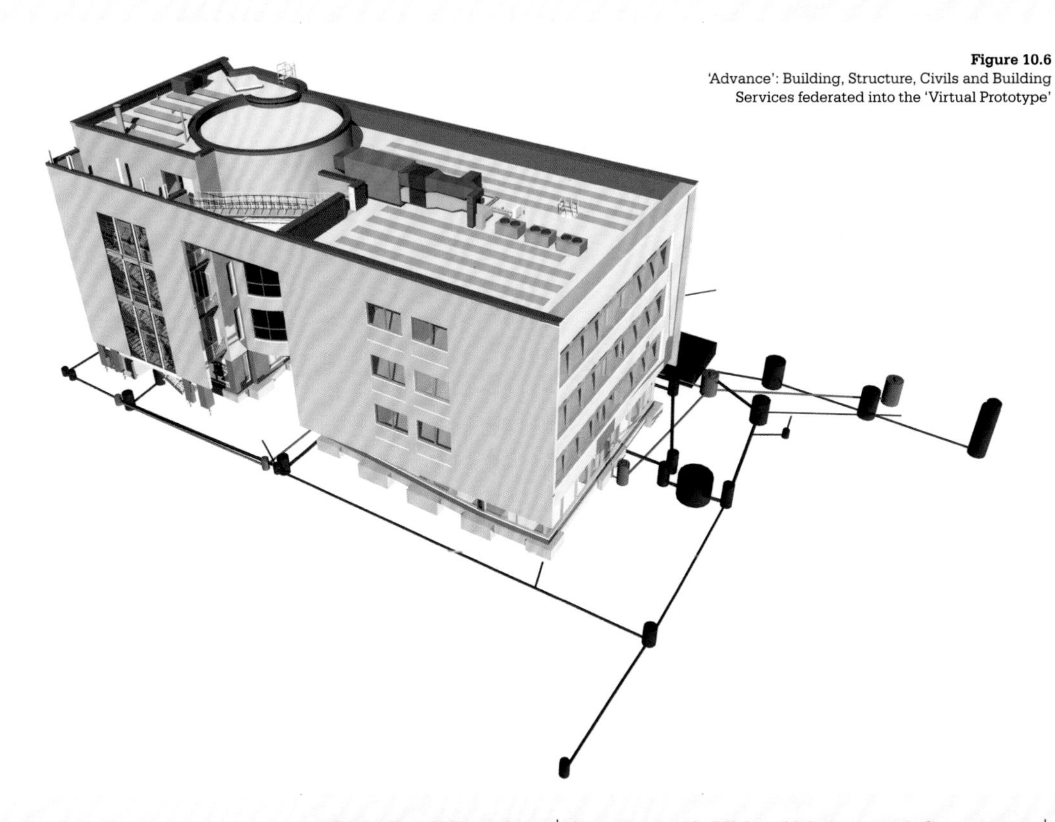

Figure 10.6
'Advance': Building, Structure, Civils and Building Services federated into the 'Virtual Prototype'

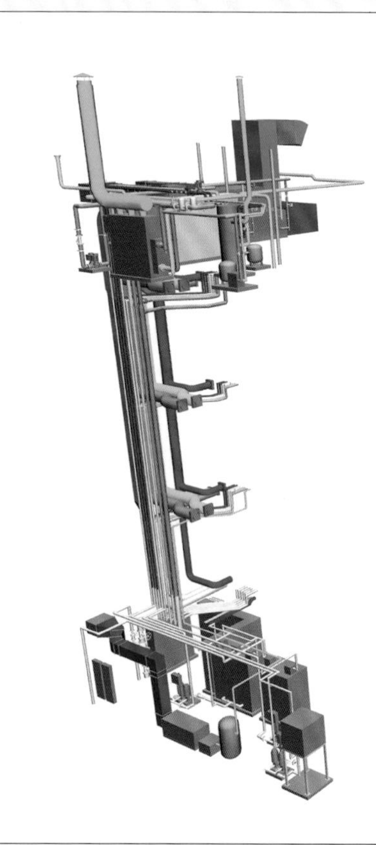

'For the first time on a Design and Build project we now had a fully coordinated design for architecture, structure and building services, before going out to tender, eliminating potential issues on-site'

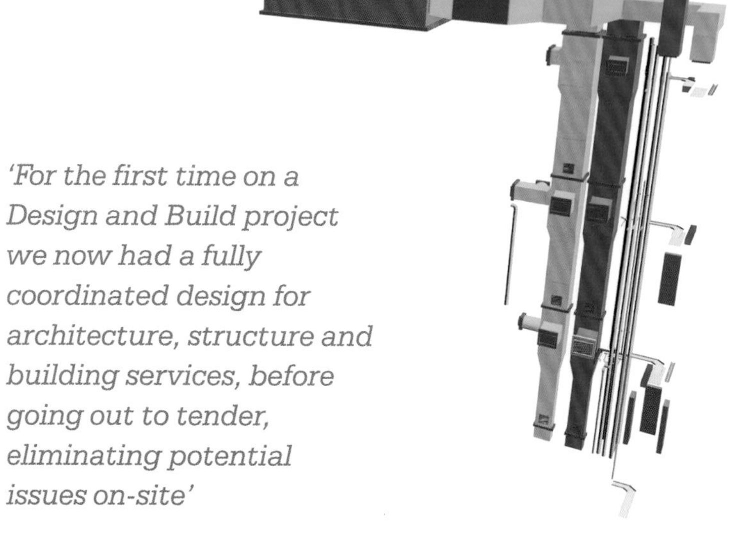

Figure 10.7
'Advance': Principal Risers and Plant Rooms. Building Services modelled to RIBA Stage E and Ductwork to DW144

Adopting BIM has allowed us to mitigate coordination issues almost entirely by the development of what we have termed the coordinated 'virtual prototype' designing the building pre-tender to BSRIA Pro-Forma 4: Technical Design for building services and RIBA Stage 4, Technical Design, for building and structure. Defining the building services design to Stage 4 at the outset gives a clearly identified set of design deliverables (which even an architect can understand) which can be modelled almost to fabrication level to deliver a fully coordinated design at tender stage. This gives the client and design team confidence that what is designed actually fits together and gives the contractor enough confidence in the design to reduce his risk register and subsequent tender.

After a number of projects where building services coordination proved to be a real issue, we persuaded our client Dudley College to model the building services design from the consultant designed proposals at BSRIA Pro-Forma 4: Technical Design to fully coordinated 'virtual prototype' using a software product called CADduct (now an Autodesk product CADmep) using specified equipment levels and to ductwork standard DW144 for ventilation systems. This is a process usually reserved for post-contract detailed design and fabrication.

For the first time on a Design and Build project we now had a fully coordinated design for architecture, structure and building services, before going out to tender, eliminating potential issues on-site.

Apart from the obvious benefit of clash detection to resolve pre-construction conflicts predominantly between structure and building services, there is scope for supply chain efficiencies because the building services were modelled in a tool like CADduct, the manufacture of the ductwork (for example) can be driven directly out of the model without the fabricator having to produce their own drawings.

This process does mean that more design work is required on building services pre-contract than that which has become the norm under Design and Build. The cost to the client in this case (at about 25k) is about the same as a handful of mistakes on your average building site! The real savings are thus in time and money, by stripping out the risks of poor coordination.

▶ DEVELOPING THE 'VIRTUAL PROGRAMME'

The further development of the model through the construction phase 'as built' (by linking it to the construction programme via a simulation package such as Synchro), allows the client and design team to see, in three dimensions, the actual progress of the project at a given moment in time, rather than the spurious progress often identified on a contractor's bar chart. Added benefits include a physical record of progress at given points in time which can be used for reference if later disputes arise.

And, in terms of stakeholder engagement the benefit is the ability to make design changes before things are built, rather than changing something that doesn't work once it's in place, with the inherent redesign and rectification costs. It is now possible with cost estimating software products to link the simulation to the 'earned value' at a given moment in time to assist in monthly valuations.

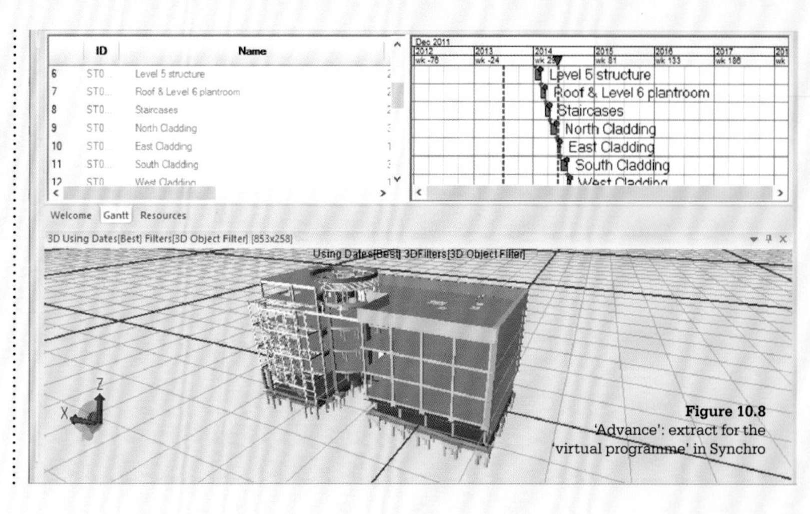

Figure 10.8
'Advance': extract for the 'virtual programme' in Synchro

CONCLUSIONS AND FUTURE PLANS

We believe there has always been a conflict of interest in a client's consultants being novated to contractors under Design and Build. Whilst the process offers some continuity, once the paymaster moves from client to contractor, by necessity so do the elegancies. In terms of the value consultants can deliver to clients, this no longer exists as they become just another subcontractor service.

Two of the perceived great benefits of Design and Build are 'guaranteed maximum price' and 'passing of all risks to the contractor'. Certainly with complex buildings, our experience of Design and Build delivers neither. There are always changes and where there are changes there is additional cost. On the question of risk, contractors 'price' risk as part of their tenders. The weaker the design and tender information, the higher the risk and the higher the price. If the risk doesn't manifest itself, the contractor keeps the money. So on both counts, guaranteed maximum price and risk, the client ends up paying more!

In terms of procurement, arguably BIM works well with traditional contracts, design teams and processes. Equally though, BIM now offers the construction industry the potential to fundamentally change the way it delivers buildings under Design and Build. If the design is delivered to full virtual prototype pre-contract with input from specialists, there is very little design work left to do after the contract has been awarded. In this case the architect could stay 'client side' post-contract and monitor the contractor's construction of the design, leaving the contractor to deliver his own 'construction' drawings. So it could be possible to have both the benefits of BIM and the benefits of Design and Build. You could call it 'Virtual Design and Build'.

We plan to develop the concept of pre-contract virtual prototyping into a 'client side' integrated professional service including architecture, structure, building services and cost planning. BIM can put 'new age' architects back, not as master builders, but as master builders of the virtual prototype. Back with the client and back in control.

There are also new ways of working together which we are now actively engaged with. The government acknowledges the deficiencies in the old forms of procurement and is rapidly promoting new collaborative ways of working such as Integrated Project Insurance (IPI). Embracing BIM, IPI is a delivery mechanism whereby all parties to the construction of a building embrace the project as a collaborative, designing with the supply chain, designing to a target cost and sharing any residual funds at the end of the project. Any overspend on the project is shared by the collaborative but capped by a single insurance policy for the whole project rather than by individuals PI cover. This eliminates the 'blame culture' inherent in design and construction as it is in everyone's interests for the project to succeed, offering true working efficiencies and guaranteed maximum price and low risk for the client.

BIM has opened up opportunities and changed the way we do business and in dark times has arguably been its saviour. For small practices, seize the opportunity and get on that jet ski! And calling all clients... if you are getting to grips with BIM, would you rather thumb a lift on the jet ski or wallow in the wake of the super-tanker?

'We plan to develop the concept of pre-contract virtual prototyping into a "client side" integrated professional service... BIM can put "new age" architects back, not as master builders, but as master builders of the virtual prototype.'

LESSONS LEARNT

- Make a decision: Small practices are adaptable and can react quickly. Remember, you are at the helm of a jet ski. There is an 18 month learning curve to get to grips with BIM so get on with it. Don't wait for a client to tell you to do it. You'll never catch up in time.
- Prepare to change the way you work: You need to learn new protocols and procedures for collaborative working so everyone in the project team is singing off the same hymn sheet. Unlike 2D CAD, there is NO margin for error. If the modelling isn't 'spot on' the whole project can fall apart.
- Adopt PAS 1192 and learn about BIM Execution Plans (BEPs) and Employers Information Requirements (EIRs). and change your preferred sub-consultants if they are not interested in joining the party.
- Don't worry too much about the capital cost. If cash is tight, think about financing the capital costs over two or three years. There is no shortage of providers who want to lend to professionals.

Part B: Contextual Chapters

01	**Why BIM?** Ray Crotty	83
02	**Getting Started with BIM** David Miller	87
03	**Tapping into the BIM community** Stefan Mordue and Rebecca De Cicco	89
04	**BIM and Diversification** Robert Klaschka	91
	BIM glossary Rob Jackson	93

WHY BIM?

Ray Crotty

01

The versatility of small practices makes the opportunity to transform the construction industry to integrated digital workflows more accessible than ever before. The new roles, new processes, better, more customised manufacturing and the huge markets of the developing world are there for the small practitioner to play a part in. This chapter discusses the fundamental changes that BIM will make to the way we design and make buildings and explores the opportunities that are open to a forward-looking and ambitious small practitioner.

The immediate practice scale benefits of BIM are if anything clearer and more easily achieved within a small practice, but the wider strategic consequences and opportunities can be missed by the smaller practitioner.

▶ The Problem with Drawings

- With drawings, no matter how detailed you get, you can only say approximately what you mean or intend.
- You depend on the person at the other end of the conversation to complete the picture, to read your drawing correctly and to understand it fully.
- And no matter how much effort you put into it, drawings will never be unambiguously clear, fully complete, correct, internally consistent and coordinated with other people's related documentation. With drawings this is simply not possible.

So huge numbers of people throughout the industry spend inordinate amounts of their time checking information, guessing the true intent, getting it wrong, correcting things, making mistakes, cutting stuff out… kango hammers. All because the only way we have of designing buildings is with drawings – drawings that produce inherently untrustworthy, unintelligent, un-computable information

▶ BIM

BIM – the combination of parametric 3D modelling systems, with rigorous communications standards and effective business-to-business protocols – changes all that. Very simply, BIM can involve the creation of one or more 3D models of a building using, not lines, but complex, fully specified, apparently intelligent, apparently self-aware components, in a powerful, computerised, 3D environment.

These components can be as detailed and as accurate as you like. They (appear to) behave just like building components in the real world. Doors, windows, walls, beams, pumps, luminaires, all display properties in the modelling system which correspond exactly with those of their counterparts in the real world. (That is actually a deceptively slippery idea). Individual models or parts of models created in this way are inherently accurate and complete.

They can contain extremely high-quality information, rich with embodied intelligence and other forms of programmed behaviour.

▶ Interoperability

The second part of the BIM concept concerns the idea of interoperability: how to make sure that this wonderfully high-quality information can be shared between modelling systems and between models and other types of industry applications, such as building performance analysis and project management systems, without becoming degraded in the process.

This issue has two parts:

- First, there must exist an agreed data exchange mechanism by means of which information can be reliably imported and exported between interfacing systems.
- And second, in a world of multiple, discipline-specific models or part models, there needs to be clear agreement as to who should provide what information, to whom, at what level of development, at any given point in the project life cycle.

BIM is the term we apply to the whole combination: modelling systems, data exchange standards and information interchange protocols.

What would happen if, instead of the poor-quality information generated by drawings, the information used in construction were fully trustworthy – needing no checking before use – and were readily computable, as the BIM vision promises?

How will this change the landscape of the construction industry and furthermore will small practices be able to take advantage or do they risk being snuffed out? The following discussion of three key areas which will undergo fundamental change should provide strategic focus for a small practice business plan.

'The small practitioner who is able to embrace change quickly should find themselves in a position to compete strongly with much larger practices, because of the staff training and structuring issues and because of their speed and cost at which they can adopt new ways of working.'

1. EFFECTIVE COMPETITION

The single greatest problem with our dependence on drawings lies in their use as the basis for the procurement of construction contracts.

Effective competition, in any market, requires that the customer can specify his or her requirements accurately and in such a manner that competing suppliers' proposals can be compared and evaluated transparently, on a true and accurate, like-for-like basis.

▶ Drawings and procurement

This is, almost inherently, impossible to achieve using drawing-based documentation.

Using this type of material, the scope of work can be interpreted to mean almost anything a bidder can plausibly claim it means – which means that there is no definitive definition of the scope of work of the contract. This in turn means that it is impossible to eliminate predatory bidders from the contracting process. All of the bidders for a given contract know this and know that one of their competitors may adopt a 'bid low, claim high' strategy.

They must all therefore bid as low as they dare and hope to make their profit on re-interpreted work, claims and other extras. This behaviour eliminates the possibility of effective competition for the operations components of construction contracts.

▶ Competition amongst contractors

Competition amongst contractors today is mainly about the marketing and estimating skills and commercial nerve required to win work and the claims management skills required to make money from projects won at cost, or less.

Skills in construction operations may give project teams a sense of pride and achievement, but are largely irrelevant to the survival of the firms they work for.

So contractors, with no existential imperative to innovate, avoid innovation risk and avoid investing in improved production methods.

Instead they subcontract and sub-subcontract, right down the supply chain, to the point where subsistence level, labour only subcontractors, working in gang-size firms, actually perform the bulk of the industry's work.

These organisations have neither the vision nor the wherewithal to invest in improving project delivery processes. As a result, effective competition exists only at the top and bottom ends of the construction industry: competition of ideas amongst designers; and product competition amongst manufacturers. Everyone in between competes to win projects – they do not compete to deliver them. This is a crucial, crippling distinction.

▶ New opportunities

For the small practitioner new opportunities will emerge through alliances with larger slow-moving corporates who need the investment in innovation that small firms can make.

Trustworthy, computable tender documents will transform this situation.

- With correct, coordinated, complete scope definition, bidders are compelled to compete on the basis of their ability to perform the construction work.
- Every scope line item can be linked directly with a component in the model and must therefore be priced explicitly.
- Every price can be compared automatically and challenged as appropriate. There are no claims opportunities, so bidders must get it right going in.

The small practitioner who is able to embrace change quickly should find themselves in a position to compete strongly with much larger practices, because of the staff training and structuring issues and because of their speed and cost at which they can adopt new ways of working.

- Predatory bidding will be eliminated.
- Contractors will be compelled to compete directly with each other on the basis of the efficiency and productivity of their project delivery techniques.

- As in other efficient markets, competition will force contractors to improve these continuously.
- Efficient firms will profit greatly – they're no longer going to be undercut by claims-hunting predators. Profit will no longer be squeezed out of the industry.
- Construction as a whole will become wealthier, able at last to invest seriously in people, methods and physical capital.

2. MANUFACTURED BUILDINGS

The precision and computability of model-based designs enable physical components of buildings to be machine-made directly, using the data contained in the modelling systems.

- The idea of ¼" and greater construction 'tolerance' will disappear.
- Individual objects will be manufactured with effectively perfect precision (where this is appropriate) and will be pre-assembled, equally precisely, in the factory, before being shipped to site.
- There, individual components and sub-assemblies will be dropped, or clipped, or slotted into place, perhaps using the sort of 'click-lock' coupling techniques used in electronics and other areas of manufacturing.

The key point is that no manufacturing from raw materials and no shaping operations – no pouring, cutting, routing, drilling, bending, folding of components – will take place on-site. The site becomes an assembly plant – comparable to a car assembly plant; the need for large numbers of skilled craftsmen and their helpers will disappear.

▶ Super-fast-track industry

It will also be a super-fast-track industry. Knowing that the other elements of the building have been (can only be) assembled exactly as designed, means that, instead of having to wait to check whether earlier elements have been built correctly, the manufacture of all components could, if required, commence simultaneously and proceed in parallel, as soon as the model has been completed. This has many advantages including competitive sourcing from whatever part of the world is most appropriate. It also hugely compresses the time required for site construction.

▶ Long guarantees

Just as long guarantees are an important attraction to buyers of motor cars and other complex products, so long guarantees are likely to drive the market for buildings of the future.

- Suppliers will emerge who will offer, say, 20 year guarantees. These will cover all of the performance characteristics of a building that can be simulated and tested in a BIM environment.
- This will include the maintenance performance of the fabric of the building and of the equipment within it, the building's energy performance and even the ease with which it can be reconfigured for new uses during its lifetime.
- The suppliers of these buildings, such as Rolls Royce with its aero engines, or closer to home, the Otis Elevator Company, will aim to derive as much of their revenues from servicing the product in its life in use, as from the initial sale.

This mode of operation will ensure that buildings of the future will be designed and built to optimise their whole life costs. It will also require that performance feedback loops become an integral part of the operation and maintenance of future buildings, ensuring that their suppliers become real learning organisations, with an inescapable commitment to the ongoing maintenance and operation of their products.

▶ Specialist subcontractors

- Specialist subcontractors may continue, but only as part of larger contracting organisations, or as licensed, thus guaranteed, installers of manufactured equipment or systems.
- Major equipment and building systems manufacturers (GE, United Technologies, Siemens, Permasteelisa…) will perhaps become main contractors or assemblers.
- They already know how product design, component sourcing and assembly processes work, as this is how they currently deal with their mainstream products: lifts, cladding systems, switchgear, mechanical equipment, etc.
- Some of these firms will surely extend the range of their activities, perhaps by acquiring construction firms, to include the production and servicing of guaranteed buildings.

Some of the opportunities of this new world of manufactured buildings are alluded to by Rob Annable in the case study by Axis Architects to achieve performance and speed of build.

The potential advantages even on smaller projects, where the architect fills the role of designer and project manager, to form alliances with manufacturers of superstructure and cladding systems is plain.

As Rob demonstrated, to a large extent these possibilities already exist, even if they are currently for building systems instead of whole buildings.

3. CHANGING ROLES IN CONSTRUCTION

In the scenario envisaged here, buildings become products; huge, complex products admittedly, but products nonetheless. And the market for buildings will be characterised by powerful, effective competitive forces which will compel their suppliers to innovate continuously.

In the wider economy, it is expected that all three of the classic factors of production: land, labour and capital, will continue to be critically scarce. The survival of firms using them will require that they be combined with the finest possible judgement.

However, the new factor of production – computing power – will become dominant; and as computers are expected to continue to follow Moore's law for the foreseeable future, so they will continue to become cheaper and cheaper over time. The future we face is one of effectively perfect information, together with effectively unlimited computing power.

▶ Programming

One result is that anything that can be programmed, will be programmed. A huge range of human activities will be subjected to this treatment and will become embedded as code in the hardware or software of computerised systems.

Most other engineering-based, rules-based, industries have already been changed out of recognition by this process. Construction will undergo a similar transformation.

For example, highly 'intelligent' programmable parametric components in modelling systems will embed their own engineering rules, so that pretty well all of the performance analysis – structural, thermal, acoustic, etc. – of almost all components, assemblies and building spaces will be carried out in BIM, as part of the architectural design of the building. Very little conventional building engineering will be required.

The benefit of this type of working environment to a small practice is huge, because the same resource is available as to the larger practices, but without the risk of implementation across a large workforce.

Building design will divide into two main areas of activity: conventional 'architectural' design, essentially the product design of complete buildings, a large part of which will involve the selection of existing components from product manufacturers' websites.

▶ Design of new components

A second part of the role will involve the design of particular new components, in collaboration with appropriate manufacturers.

Designers will have to become as familiar with products, their interfaces with each other and their construction requirements as classical architects used to be with the materials that were available to them; a new profession of digital master builders (archi-tecton), perhaps.

This may seem a far-fetched and unlikely development of the architect's role. But it's essentially how many modern stadium buildings are designed and constructed and it's largely how Frank Gehry's studio has worked for the past 20 years.

Construction is the largest remaining source of employment for very large numbers of men of low levels of education: craftsmen and labourers. It also employs large numbers of people who carry out routine, rules-based work: code engineers and administrators and, of course, the armies of checkers, man-for-man markers and so on.

▶ Digitisation of construction

The digitisation of construction will follow the same general pattern as was experienced in other engineering industries – and massive structural reductions in employment levels will follow. The timing of the transition from analogue to digital in construction is difficult to predict accurately, but when it happens, it will be brutal and sudden.

This is part of the process described by the American economist David Autor[1], amongst others, as the progressive 'hollowing out; polarisation of society' that has been observed in the USA and other developed economies since the 1980s.

However, bear in mind what Albert Einstein said (it's always worth bearing in mind what Einstein said): 'Imagination is more important than knowledge. For knowledge is limited to all we now know and understand, while imagination embraces the entire world and all there ever will be to know and understand.'

Thus, knowledge can be codified, but imagination cannot. Human attributes such as empathy, courage, leadership and the thing we call vision are all fundamentally like imagination; it is difficult to imagine how they might be computerised. These human characteristics are now and will remain, central to the genius of great design and construction.

4. CONCLUSION

In conclusion, notice that each of the issues discussed here combines threats and opportunities in ways that have never been seen before. Remember too, that we tend to over-stress the short-term impacts of new technologies and to greatly under-stress their long-term effects.

To survive in this world SMEs will need to become and continue to be, aware of global developments in the construction industry but also to look outside construction and keep abreast of information-driven innovation in other industries. Peoples' firms and their jobs will still be changed beyond their wildest expectations, but if they stay alert, at least they'll have a chance of riding the wave rather than being swamped by it.

For the small practice with low risk aversion the list of opportunities is long and the future bright for those who can transform themselves. You could argue that the changes that are to come are well suited to the architect's mindset, with the obsession for reinvention, exploration, complexity and discovery. One thing is for sure – there will be winners and losers in this bright future.

[1] For a summary of his ideas see: *http://opinionator.blogs.nytimes.com/2013/08/24/how-technology-wrecks-the-middle-class/?_r=0* [Retrieved 26 September 2013]

GETTING STARTED WITH BIM

David Miller

02

As a small practice, there is clear evidence that BIM adoption has underpinned the year on year growth that David Miller Architects (DMA) has achieved in the past five years. This chapter describes practical steps to help newcomers get started and explains why new technology, a difficult economic environment and the clear focus from UK government have created an opportunity for small practices to change the odds through BIM.

1. WHAT IS THE MOTIVATION FOR ADOPTING BIM?

We believe that there are currently two drivers for BIM adoption, both of which will work, but that adopting BIM 'just because you have to' is missing the real opportunity.

- We are all aware of HM Government's mandate and we are beginning to see BIM included in PQQs.
- But having been through the process we believe there are opportunities that make BIM adoption much more attractive to small practices. These include increased efficiency and profitability and a better way of working.

We know that as designers we add considerable value through our ideas; however the time spent developing these ideas eats away at our fee. What we quickly realised was that BIM tools were offering us speed, accuracy, improved coordination and quality control – all helping to protect our fee.

We also realised that as a small practice we had it easy when it came to rolling out BIM, because the real barrier isn't cost, but change management. As a smaller organisation we were able to spot the opportunity and make the change very quickly. We were able to act on experience and instinct, without having to convince boards or middle managers. Larger organisations simply don't have this agility, therefore BIM is a real opportunity for small practices to capture market share.

2. THE ADOPTION PROCESS

Roll-out is incremental as there is a learning curve. At DMA we have been getting steadily more sophisticated over the last four years. However I wish that we were starting now, as this is the optimal moment. The process that we have been through can now be collapsed into just months given the clarity of the government mandation and the wealth of information and support that is now available. Being able to start out on smaller projects is another advantage for small practices. We had the safety net of knowing that we could have easily reversed out of our early projects into CAD if we had got into difficulty.

However the most important moment is making the commitment and for us this was appointing a BIM Champion to lead the process. Our BIM Champion developed a six-day training schedule broken down into 40-minute modules so it was easier for the team to find time to do a module. She then introduced BIM Boot Camp for new starters, which involves full immersion in their first week in the office before there are any project distractions.

Once BIM is embedded into your office culture you will find that individual members of the team start to take responsibility to explore different parts of the process and share their experiences. This led us to broaden the services we are able to offer to include rapid energy modelling, rapid cost modelling and logistics planning. We're currently exploring opportunities around facilities management.

3. COST CONSIDERATIONS

For DMA it's been around £10,000 per workstation, so BIM is a big investment. This figure includes hardware, software and training. Remember though that the costs for bigger practices are the same, just scaled.

We have been spending around £30k a year over the last four years. Whilst this felt like a lot early on when there were only four of us, now it is far less painful to add on new workstations as we go along and this confirms that there is a financial push at the beginning.

It's important to note though, that half of that cost is training. When you view cost in relation to salaries and fee income it looks far less alarming and it is easy to see why we view BIM as an investment in our team and working methods rather than in technology.

4. SUPPORT

The framework that brings everything together and which has given the UK construction industry such a clear focus is the government construction strategy. This has been elaborated and built upon by the BIM Task Group and it's fair to say that the Task Group website (www.bimtaskgroup.org) is the 'go to' location for reliable information. If the model is the single source of truth in a project, then the BIM Task Group website could be seen as the single source of truth for BIM adoption.

The site provides:

- Contract documentation: such as the BIM Protocol which defines the contractual arrangements for a BIM project, making collaborative working a contractual requirement and therefore realistically possible.
- It answers the commonly asked questions around copyright and intellectual property, especially with regard to sharing model data.
- There are Employer's Information Requirements which clarifies to the team what to include in the model and to what level of detail.
- There is the new Plan of Work, which will give consistency across disciplines. For BIM to work we do need to be doing certain things at the same time.
- And there is clarification and examples of COBie outputs and implications on Government Soft Landings, also mandated for 2016.
- Hugely useful is PAS 1192 Part 2. This gives a standard for sharing information between disciplines during the Design and Construct phase of a project. Part 3 is due to follow, which will cover the Operation phase.

There are also links to sector-related groups including BIM4SMEs. These groups are there to support the industry by disseminating information around the government's mandate and, equally importantly, to act as a feedback mechanism to the Task Group, so industry can be heard. This is further supported geographically by the Construction Industry Council's BIM regional hubs whose role is similar, but not sector specific.

The best thing is that all of this is free! So if you read the Construction Strategy and make the BIM Task Group website your home page, you are off to a good start!

5. WHAT IS THE BENEFIT?

From our point of view, BIM has made us more efficient and we're sure it has given us a competitive advantage. Being able to offer additional services has differentiated us in a difficult market. We are also confident that the consistency in our output has generated repeat business, not least because we have been able to drive efficiencies from project to project as our database of components has become richer and more refined.

The new processes are encouraging collaboration at every level and re-focusing the team on the end product rather than just project trackers and compliance reports. The virtual building is on our monitors and the office projector screens constantly, so it's a much more satisfying way to work.

The quality controlled output has also given our team confidence; BIM tools have reduced the drudgery of production of information, enabling the team to punch above their weight, which again has improved morale.

Importantly, stakeholders can see exactly what they are going to get, which is managing their expectations and smoothing the projects through the review and approval process. This is happening at every level of the project from concept design through to safety on-site.

As a bonus we found that our new workflow has simplified design management and allowed us to build our internal processes around the BIM. All of our external accreditations including ISO9001 Quality Management and ISO14001 Environmental Management have been positively affected by BIM.

6. CONCLUSION

In terms of the bottom line, reviewing fee income against technical costs, the impact on business is clear. There is a period of adjustment, but once embedded there are real efficiencies to be gained. Some of this could be a consequence of a small practice growing and taking on larger projects, but that in itself could be a consequence of using BIM tools.

We believe BIM has allowed us to grow in a difficult market and to take on bigger and more challenging projects. This is why we firmly believe that BIM is a great opportunity for small practices!

TAPPING INTO THE BIM COMMUNITY

Stefan Mordue and Rebecca de Cicco

03

With much focus on the public sector and large-scale projects, it is often a misconception that BIM is only the reserve of cash-rich organisations undertaking vast and complex schemes.

On the contrary, it is perhaps the small, nimble and adaptable practice that has much to gain from what BIM has to offer. The real barrier is change management and how we, as humans, adapt. It is here that the small practice gains the upper hand.

1. COMMUNICATION

As we progress in this digital age, communication and interaction via social networks such as Twitter, Facebook and LinkedIn is on the increase. Through the power of tweets, status updates, blog posts and online forums the small practice has the ability to tap into a world of online communities where there is a real sense of passion and enthusiasm and an environment of collaboration and knowledge sharing.

Individuals should be able to access this information easily and influence a team without having to focus on practice strategies or needing to force their ideas through differing levels of management.

It is here that there is an opportunity for the SME to gain an online presence within the BIM environment as an agent of change.

▶ Community ideology

Whether it's a link to a document, a discussion on best practice or organising a tweet, the online community provides a vast untapped free resource of information and comradeship. This ideology has been helped by various leading industry figures showcasing and sharing their knowledge to those around them. Events such as BIM Show Live only began as small entities within the community and have now grown and expanded to become industry leading events.

▶ Communicating and networking

Online communication must be accompanied by physical meetings and it is important to recognise that the aim of these incentives is to meet others involved in the field to share knowledge.

Networking is hugely important, allowing those interested in BIM to meet regularly to discuss themes and the progression of BIM in the industry. It is here that we have seen the success of the Construction Industry Council's regional BIM Hubs, whose aims are to up the skillset of the industry and be present and vocal across the country.

The Task Group brings expertise from industry, government, public sector, institutes and academia and produces a monthly newsletter outlining forthcoming events and developments across the UK. As well as providing a good clear set of frequently asked questions the site offers information on a series of BIM processes, as well as lessons learnt from the government trial projects.

▶ Partner organisations

Central to the BIM Task Group are a number of partner organisations, including the BIM4 groups which represent various areas of the industry.

A specific BIM4SMEs group is made up of individual organisations from different sector backgrounds including specialist subcontractors, architects, FM providers and construction bodies. All these organisations are SMEs or have a key interest in ensuring SMEs are well informed and recognised for their advancing of BIM.

2. KEY RESOURCES

The resources below outline current industry thinking. They include events, online communities and individuals influencing BIM. These resources are growing by the day as more BIM users create their own voices.

▶ BIM Task Group

- www.bimtaskgroup.co.uk
- @BIMgcs

The group's aim is to support the government construction strategy's main objective of achieving Level 2 BIM by 2016. The site provides support, access to all BIM4 Groups and lessons learnt documents. It highlights the latest news and quick industry updates. The website also contains the latest links to the government projects.

▶ Construction Industry Council and Regional BIM Hubs

- www.cic.org.uk
- www.bimtaskgroup.co.uk
- @BIMHubs
- @CICtweet
- @CICCEO (Graham Watts)
- @BIM2050

The CIC set up the Regional BIM Hubs with the BIM Task Group to encourage the industry in BIM adoption. They are authors of the CIC BIM protocol, at the heart of the legal framework of the 1192 suite of standards. The Hubs run free events across the UK and their primary aim is to share knowledge throughout the industry and disseminate information generated by the BIM Task Group around its incentives and goals leading to 2016.

▶ BuildingSMART UK

- www.buildingsmart.org.uk

BuildingSMART's primary aim is to improve process and training across the UK. They work with leading figures across differing sectors to define standards. BuildingSMART events and workshops are worthwhile to those wanting to adapt to new technologies and ideals relating to BIM implementation.

▶ BIM Diary

- www.bimopedia.com
- @BIMdiary

Follow for the calendar-listed entries of most BIM events in the UK.

▶ Twitter

- #ukbimcrew
- various BIM related Tweeters

The #ukbimcrew has become a legacy to the early adopters in the UK seeking out and sharing knowledge amongst each other on an open platform. The group is not exclusive, nor does it focus on one area of BIM implementation or process. It is rather a search mechanism for the industry to seek out an exhaustive array of discussions, debates, questions and information all relating to the UK BIM community. It is a collection based on multiple users experiences and has been collated from late 2011.

▶ National BIM Library

- www.nationalbimlibrary.com

The National BIM Library, in conjunction with the NBS, has developed a library of BIM-based components for the industry.

For those on their early BIM journey the NBL helps to support BIM-enabled processes and provides for both generic and manufacturer's specific content for building information models. The NBS also supports Dr Stephen Hamil's blog (Director of Design and Innovation at RIBA Enterprises) which is a useful resource for up and coming information relating to BIM:
www.constructioncode.blogspot.co.uk
@StephenHamilNBS

▶ BIM Show Live

- www.bimshowlive.co.uk
- @BIM_UBM

This is an annual event that draws in all industry leading figures to discuss their project experiences in relation to BIM. It is an important event to not only learn from those leading the industry, but also to network.

▶ The Construction Project Information Committee (CPIC)

- www.cpic.co.uk

CPIC are the focus of a number of well known BIM related projects including Uniclass 1.4 and the energing Uniclass 2 standards for classification of information. They also provide a number of useful documents that are used in the pre-qualification process for Level 2 BIM projects.

▶ ThinkBIM

- www.ckehub.org
- @thinkbim

ThinkBIM is a multidisciplinary group from Leeds Metropolitan University passionate about industry change, collaboration and sharing knowledge. The team consists of a variety of industry professionals all passionately involved in aiding the industry in its development of BIM-enabled processes. They regularly host events and ensure all disciplines are addressed.

▶ Software User Group Events

- www.gbrugs.co.uk,
- www.benteyuser.org

Most software vendors support user groups globally and both Autodesk and Bentley have a series of user groups spread across the country and indeed across the world. The groups help to develop technical solutions for a cross-disciplinary approach to learning and the events enable sharing of real life solutions when it comes to BIM authoring software solutions.

▶ BIM Small Practice Perspective Events

RIBA hosted a series of events in 2013 and 2014, as an introduction to BIM implementation from the viewpoint of the SME, focusing on sharing knowledge for architects in the UK. Future events will be a good resource for SMEs to make connections to those early adopters pushing forward in technological solutions for building and design.

3. CONCLUSION

The UK community is moving incredibly quickly in its adoption of BIM-enabled processes and it is important for the SME to remain agile and open to developments.

The resources will continue to evolve and change as we become increasingly more sophisticated and therefore it is important to stay informed and keep abreast of new updates and individuals involved in BIM.

BIM AND DIVERSIFICATION

Robert Klaschka

04

Many small practices who have taken the plunge with BIM and prioritised investment in technology find that their initial ideas, motivated by improving internal productivity or client engagement, actually open up entirely new streams of business.

In this book, five of the authors refer to new business opportunities that have sprung from embracing new technologies. SMEs are ideally placed to take advantage of these because many of the opportunities exist as a result of the catch up that the industry in general is playing with the implementation of BIM.

Thus, for example, many architectural practices who are trying to get to grips with BIM need to outsource these tasks or need training and advice in how to get started themselves. For some BIM-ready SMEs this has changed the way they do business and, in some cases, the focus of their work has moved beyond the boundaries of architecture.

▶ Consultancy and training

Consultancy and training are two of the first and most straightforward opportunities for extending the services that architects find they can offer when they have started to master BIM in their own offices. In many instances this may have sprung from the desire to get other consultants working collaboratively so that all can benefit.

▶ Vectorworks

Jonathan Reeves Architects, as you can see from Case Study 02, has gone down the route of providing consultancy to other architects to help them develop their processes and understand what will be asked of them.

This has led onto providing training resources for Vectorworks, an entry-level, low-cost BIM package, to the same practices. Focusing on the software JRA use has meant that the architects and engineers he is providing training and advice for are working on similar sized projects.

▶ Revit

Similar to JRA in many ways Kara de los Reyes from Little BIM Studio Ltd, a Bristol-based micro practice started in 2011, started out providing architectural consultancy work but soon recognised a gap in the market.

The ambitious government target of achieving Level 2 BIM maturity by 2016 was a path that some would struggle with if there wasn't assistance. Her subsequent success in providing Revit training and modelling demonstrates the need for professional services from someone who really understands what is important in a BIM workflow.

Little BIM has capitalised on both the skills shortage and the considerable hardware and software costs required to work using BIM software. Having worked with several practices who had invested in software but were using old hardware she was surprised to find just how slow operations could be on outdated machines. For example it could take 30 seconds renaming a sheet and as long as six minutes to open a file.

By initially helping with technology issues she has developed a symbiotic relationship with the practices she is working with that has expanded to advising about the significant changes to programme expectations and workflows that a design practice will need to make to fully embrace BIM.

▶ Existing BIM Services

In my case study I discuss how we have become specialists in the use of point clouds to capture existing buildings that we were refurbishing. This has led onto our own businesses diversification to provide models of existing projects for other design teams, mainly contractors. Compared to architectural projects the work is relatively predictable because it has a clear procedure and a defined scope and can't be delayed by some of the more subjective issues that can hit the design process when a client doesn't like the appearance of what you have designed, or the brief changes.

Also because there are currently relatively few companies providing existing BIM in an environment with a high demand it's possible to negotiate cost. The effect of this is that the practice has a much more predictable baseline cash-flow than architectural work alone can provide.

Another benefit of this model is that it allows us to demonstrate our ability to work on larger projects without the risk of a first time design opportunity. For example, we recently delivered a building information model of an existing school to a main contractor, which has enabled the practice to demonstrate its ability as a high-quality information provider and has established us as a potential designer for future projects.

They want to have the quality of information we are providing on new-build work.

▶ Content creation for manufacturers

One of the biggest issues in BIM is where to source the components to build your models from.

BIM applications are delivered with generic objects (walls, doors, windows, etc.), but there is a demand for manufacturer specific objects.

A number of companies have started to provide a service to create content for manufacturers but often they are platform specific. Metz (Case Study 10) in addition to providing conventional architectural services, have a long-standing relationship with Howitt Consulting, a company specialising in relationships and development offerings for construction product suppliers.

Working together they had realised that there was not only going to be a supply shortage for BIM 'objects', or Content as it is now known, but also a knowledge shortage of what BIM is all about within the supply chain.

They have formed a joint venture company SparkIFC to deliver cross-platform content to product suppliers and manufacturers.

▶ Software design

Other architects have found that, in embracing technology, they have branched out into software design and development. That is exactly what architects Michael Kohn and Renee Puseep did with their company Slider Studio.

BIM is fundamentally a data-based process. The difference with Slider Studio was that the data they wanted to share was not handled by the market leading BIM software applications. They have understandably tried to distance themselves from the term BIM because of its association with the small number of software applications that we have come to think of as defining it.

Nevertheless, Slider Studio develops its own software for managing both spatial and non-graphical information. These are mainly focused on the front end of the design process and often facilitate community engagement with design to encourage integration of user feedback.

For example, Stickyworld is a community-facing web-based tool designed to provide information to the community on local building developments and collect their feedback.

While this may seem far from the scope of most BIM applications there is no question that what they do facilitates the mass dissemination, collection and aggregation of information about both buildings and projects. I'd call that BIM.

▶ Other methods of diversification

Beyond these examples there are an increasing number of other roles that an architect could move into. Projects need BIM managers or coordinators and validation of data.

Clients need advice on the BIM process as well as establishing and documenting their own standards and deliverables. For example what should be modelled at each stage of a project and what data should be attached to it.

Small innovative practices who are ahead of the game are ideally placed to provide these and other emerging services. Diversification offers the opportunity to insulate your business from some of the vagaries of the architectural profession such as aggressive fee competition and reliance upon unstable work pipelines.

Often providing the type of services discussed in this chapter is more predictable and, in many instances, is perceived as higher value than design work. You are also likely to have a much smaller pool of competitors, who are likely to approach problems differently than a designer would.

▶ Not for the faint-hearted

However diversification is not for the faint-hearted. Moving into unknown territory has its hazards and needs careful consideration. What are the contractual consequences of data you are providing? Are the additional services you are providing covered by your professional indemnity insurance?

CONCLUSION

Young practices with their networks and savvy social media campaigning are ideally placed to have both the vision and the contacts to make unique offerings as the examples above demonstrate. To diversify you will need resilience, perseverance, skilled assistance, self-belief and possibly some luck.

Yet consider what makes a successful young practice above design ability and also consider that the broad nature of the architectural education and mindset has historically led to architects working across a much wider field. The architectural mindset is one that naturally lends itself to diversification.

BIM glossary

This glossary is abridged and edited from **bimblog.bondbryan.com** produced by Bond Bryan and maintained by Rob Jackson

All terms marked * are as defined by PAS 1192-2:2013

All terms marked ** are as defined by DRAFT PAS 1192-3:2014

Analysis the action or process of analysing the model(s) for different purposes or a table or statement of the results of analysis of the model(s)

Assembly a composition or collection of components and/or modelled elements arranged to define part or all of a building, model, structure or site. An assembly typically contains information that can be referenced without repositioning a group of components or types to enable the reuse of standardised design or specification elements improving productivity of design and delivery as well as providing a location to hold specifications and lessons learnt in a simple and useable way. They may hold benchmark data for cost and carbon impacts. The contents of assemblies may themselves have attributes and classifications. These properties may include key data which is attached (to the object) for use once it is placed into a model and may include cost, CO_2, programme, maintenance and other key information *

Asset Information Model (AIM) a maintained information model used to manage, maintain and operate the asset *
OR
structured and unstructured data and information that relates to assets to a level required to support an organisation's asset management system. An AIM can relate to a single asset, a system of assets or the entire asset portfolio of an organisation **

Asset Information Requirements (AIR) data and information requirements of the organisation in relation to the asset(s) **

Attribute a piece of data forming a partial description of an object or entity *

Author the originator of model files, drawings or documents *

Bill of Quantities (BQ) a list of items giving detail identifying descriptions and firm quantities of the work comprised in a contract

BIM Execution Plan (BEP) a plan prepared by the suppliers to explain how the information modelling aspects of a project will be carried out *

'BIM wash' a term describing the inflated – and sometimes deceptive – claim of using or delivering BIM products or services

CIC scope of services multidisciplinary scope of services published by the Construction Industry Council (CIC) for use by members of the project team on major projects *

Clash detection detecting possible collisions between elements in a building information model which would not otherwise be desired or buildable on site

Classification a systematic arrangement of headings and sub-headings for aspects of construction work including the nature of assets, construction elements, systems and products *

COBie (Construction Operation Building information exchange) structured facility information for the commissioning, operation and maintenance of a project, often in a neutral spreadsheet format that will be used to supply data to the employer or operator to populate decision-making tools, facilities management and asset management systems *
OR
structured asset information for the commissioning, operation and maintenance of an asset often in a neutral spreadsheet format that will be used to supply data to the owner or operator to populate decision-making tools and asset management systems **

Common coordinates a way of identifying the location of the model(s) or building(s) in relation to a specific agreed point. This point could be to global or local coordinates and this should be identified as to which has been used

Common Data Environment (CDE) a single source of information for any given project, used to collect, manage and disseminate all relevant approved project documents for multidisciplinary teams in a managed process *
OR
a single source of information for the project which collects, manages and disseminates relevant approved documents relating to the project (as defined by the CIC Outline Scope of Services for the Role of Information Management)
OR
a single source of information for any given project or asset, used to collect, manage and disseminate all relevant approved project documents for multidisciplinary teams in a managed process **

Component an individual building element that can be reused. Examples include doors, stair cores, furniture or internal room layouts, facade panels, etc. Components are typically inserted and moved/rotated into the required position, a synonym for 'occurrence' *

Computer-Aided Facilities Management (CAFM) the support of facility management by information technology

Constraint a mathematical expression, often algebraic, defining equalities (=) or inequalities (>,<) across various parameters. Constraint may be geometrical, such as parallelism, or a specified angle relation

Container named persistent set of data within a file system or application data storage hierarchy including, but not limited to, directory, sub-directory, data file, or distinct sub-set of a data file, such as a chapter or section, layers or symbol (as defined by BS1192:2007 - 3.2)

Data capture putting information into a form that can be fed directly into a computer

Deliverables the specific requirements for the project which may be generated directly from the model or from other sources. They may include the building information model, drawings, fly-throughs, images, data, schedules or reports

Design intent model the initial version of the Project Information Model (PIM) developed by the design suppliers *

Design team the architect(s), engineer(s) and technology specialists responsible for the conceptual design aspects of a building, structure or facility and their development into models, drawings, specifications and instructions required for construction and associated processes. The design team is part of the project team

Document a container for persistent information that can be managed and interchanged as a unit (as defined by BS1192:2007 - 3.4)
OR
information for the use in the briefing, design, construction, operation, maintenance or decommissioning of a construction project, including but not limited to correspondence, drawings, schedules, specifications, calculations, spreadsheets *

Drawing a document used to present graphic information (as defined by BS1192:2007 - 3.5)
OR
static, printed, graphical representation of part or all of a project or asset *

Employer the individual or organisation named in an appointment or building contract as the employer *
OR
the person appointing the project team member pursuant to the agreement and any valid assignee of the employer's rights and obligations under this protocol subject to the terms of such assignment

Employer's Information Requirements (EIR) a document setting out the information to be delivered, and the standards and processes to be adopted by the supplier as part of the project delivery process *

Element a construction entity part which, in itself or in combination with other such parts, fulfils a predominating function of the construction entity (as defined by ISO 12006-2:2001 - 2.7)

Energy analysis the action or process of analysing the model(s) from an energy point of view or a table or statement of the results of analysis of the model(s)

Facility management management during the operational phase of a facility or building's life cycle, which normally extends over many decades. It represents a continuous process of service provision to support the client's core business and one where improvement is sought on a continuous basis

Federated model a model consisting of linked but distinct component models. These can be drawings that do not lose their identity or integrity by being linked, so that a change to one component model does not create a change in another component model

Field part of a container name reserved for meta-data (as defined by BS1192:2007 - 3.6)

Graphical data data conveyed using shape and arrangement in space *

Industry Foundation Classes (IFC) is a neutral and open specification that is not controlled by a single vendor or group of vendors. It is an object-based file format with a data model developed by BuildingSMART to facilitate interoperability in the building industry

Information management measures that protect and defend information and information systems with respect to their availability, integrity, authentication, confidentiality, and nonrepudiation. These measures include providing for restoration of information systems by incorporating protection, detection, and reaction capabilities are tasks and procedures applied to inputting, processing and generation activities to ensure accuracy and integrity of information *

Instance an occurrence of an entity at a particular location and orientation within a model (as defined by BS1192:2007 - 3.7)

Interoperability the ability of two or more systems or components to exchange information and to the use the information that has been exchanged

Integrated Project Delivery (IPD) a collaborative alliance of people, systems, business structures and practices into a process that harnesses the talents and insights of all participants to optimise project results, increase value to the owner, reduce waste, and maximise efficiency through all phases of design, fabrication and construction

Laser scanning controlled steering of laser beams followed by a distance measurement at every pointing direction used to rapidly capture shapes of objects, structures, buildings and landscapes

Layer a container comprising selected entities, typically used to group for purposes of selective display, printing and management operations (as defined by BS1192:2007 - 3.8)

Lean principles understanding value from a client's perspective, identifying the value stream, establishing a balanced flow of work, in which the demand for product is pulled from the next customer, with a constant drive for continuous improvement and perfection (based on 'Lean Thinking', Womack & Jones, 2003 edition) *

Levels of definition the collective term used for and including 'level of model detail' and the 'level of information detail' *

Level of detail the level of detail required for a Model (CIC BIM Protocol, first edition, 2013)

'Lonely BIM' a non-collaborative 3D model produced by a single designer

Life cycle assessment the methodology used to establish a life cycle costing for a material or a product

Master Information Delivery Plan (MIDP) the primary plan for when project information is to be prepared, by whom and using what protocols and procedures, incorporating all relevant task information delivery plans *

Meta-data data used for the description and management of documents and other containers of information (as defined by BS1192:2007 - 3.9)

Metrics acceptability of the deliverable may be assessed against the requirements shown in the examples and/or against indicative ratios and counts based on the information provided *

Model a collection of containers organised to represent the physical parts of objects, for example a building or a mechanical device (as defined by BS1192:2007 - 3.10)
OR
a three-dimensional representation in electronic format of building elements representing solid objects with true-to-scale spatial relationships and dimensions. A model may include additional information or data that reflects the physical and/or functional characteristics of the project

Object any part of the perceivable or conceivable world (as defined by ISO 12006-2:2001 - 2.1)
OR
an item having state, behaviour and unique identity – for example, a wall object *

Open BIM a unique approach to collaborative design and realisation of buildings allowing project members to participate regardless of the tools they use

Origin a setting out point for a project or programme using co-ordinate geometry or related to the OS or geospatial reference *

Originator the agent responsible for production of a container (as defined by BS1192:2007 - 3.11)
OR
the author of models, drawings and documents *

Parameters variables used in a function or equation to assign values: coordinate, dimension, material, distance, angle, colour, unit price, energy coefficient, and so forth

Placeholder simplified or generic representation of a 3D object *

Plain language questions questions asked of the supply chain by the employer to inform decision-making at key stages of an asset life cycle or project **

Point cloud a set of vertices in a three-dimensional coordinate system. These vertices are usually defined by X, Y and Z coordinates, and typically are intended to be representative of the external surface of an object. Point clouds are most often created by 3D scanners. These devices measure in an automatic way a large number of points on the surface of an object, and often output a point cloud as a data file. The point cloud represents the set of points that the device has measured

Project BIM protocol the project-specific BIM protocol setting out the obligations of the principal members of the project team in respect of the use of BIM on the project

Project information plan the plan for the structure and management and exchange of information from the Project team in the information model and the related processes and procedures

Project Implementation Plan (PIP) a statement relating to the suppliers' IT and human resources capability to deliver the EIR *

Published information refers to documents and other data from shared information. Typically this will include exported data, contract drawings, reports and specifications (reference BS1192:2007) or a component of the CDE for drawing renditions that have been approved as suitable for a specific purpose – for example, suitable for construction *

RACI indicator an abbreviation used to identify which of a group of participants or stakeholders are responsible for ('R'), authorize ('A'), contribute to ('C') or are kept informed about ('I') a project activity *

Requirements the documented expectations of facility owners/ commissioners for sharable structured information. These are also referred to as the Employer's Information Requirements (EIR) (alternatively, the Client's Information Requirements) *

Revision used to identify revisions of documents, drawing and model files *

Shared information that has been checked and approved and made available across the project team such as information for data exchange between BIM software, like gbXML, CIS/2 and IFC files
OR
a component of the CDE. The shared section of the CDE where information can be made available to others in a 'safe' environment. The early release of information assists in the rapid development of the design solution. To allow this to be achieved, the concept of information 'status/suitability' has been adopted *

Soft landings graduated handover of a built asset from the design and construction team to allow structured familiarisation of systems and components and fine-tuning of controls and other building management systems *

Status defines the 'suitability' of information in a model, drawing or document. Not to be confused with the status in architectural documentation as 'new-build','retain' or 'demolish' *

Supplier Information Modelling Assessment Form a form conveying the capability and experience of a supplier to carry out information modelling in a collaborative environment *

Supplier Information Technology Assessment Form a form conveying the capability and IT resources of a supplier for exchanging information in a collaborative environment *

Task Information Delivery Plan (TIDP) federated lists of information deliverables by each task, including format, date and responsibilities *

Uniclass (Unified Classification for the Construction Industry) published by the Construction Project Information Committee (CPIC), this is a UK standard for classification

Version sub-indexing to revision as used in the common data environment to show the development of information and information models, e.g. if a version is named P1.1, P1 is the revision number and the .1 is the version to that revision *

Work In Progress (WIP) each individual company or discipline's own work. This is information that has not been approved or verified as fit to share across the project team (reference BS1192:2007)

http://bimblog.bondbryan.com/wp-content/uploads/2014/01/30004-BIM-dictionary.pdf

Image credits

Architype Ltd	59, 61–64
Axis Design Architects	27, 29–32
Chris Hodson	75: top
Constructive Thinking Studio Ltd	35, 37–41
David Miller	87
Dudley College	76: top
Eurobuild	67–71
Fulcro	79
jhd Architects	19, 21–25
Jonathan Reeves Architecture	11–17
Metz Architects	73, 75: bottom left and right 76: bottom, 77, 78
Niven Architects	43, 45, 47
Poulter Architects	3–9
Ray Crotty	83
Rebecca de Cicco	89: right
Stefan Mordue	89: left
Studio Klaschka	51–57, 91